Engineered Writing

A Manual for
Scientific, Technical, and
Business Writers

Engineered Writing
Second Edition

*A Manual for
Scientific, Technical, and
Business Writers*

**By
Melba Jerry Murray and
Hugh Hay-Roe**

PennWell Books
PennWell Publishing Company
Tulsa, Oklahoma USA

Library of Congress Cataloging in Publication Data

Murray, Melba Jerry.
   Engineered writing.

   Updated ed. of: Engineered report writing.
Rev. ed. [1969]
   Bibliography.
   Includes index.
   1. Report writing.   2. English language—Rhetoric.   3. English
language—Technical English.   4. Technical writing.   5. English
language—Business English.
I. Hay-Roe, Hugh.   II. Murray, Melba Jerry.
Engineered report writing.   III. Title.
PE1478.M8   1986   808'.0666   85-17294
ISBN 0-87814-293-2

Printed in the United States of America 1991

Table of Contents

v

List of Illustrations

About This Book

As the title suggests, this book is an outgrowth of *Engineered Report Writing,* first published in 1963. Like any how-to book that stays in print long enough, the original version merited an update. During the 20 years since it introduced the concept of well-engineered writing, momentous changes in the working world have transformed the way we produce written documents and the way executives, professionals, and educators think about communication.

The computer has opened up a whole new spectrum of aids for writers. The international business scene has grown even more complex, and intercultural communication more important. Current English usage has been influenced by social changes—sometimes for the better, more often for the worse. The economics of communication is vastly different from what it was even a few years ago; costs have soared, making efficient communication more critical than ever before. Students leaving the university without good communication skills find it increasingly difficult to advance professionally, even to obtain the jobs they prefer. And finally, technological competition has intensified so drastically that effective communication has become crucial to the economic welfare of individuals, small businesses, big corporations, and educational institutions alike.

This book brings *Engineered Report Writing* into this new age. It is based on the realities we have experienced in teaching, consulting, and working in industry over a period of more than 25 years. It deals with new problems raised by corporate managements and individual on-the-job writers we have worked with. New applications of the Murray System are included. Years ago we learned that one can properly engineer not only reports, but also business letters and memorandums; scientific articles and professional papers; technical newsletters; speech texts; agenda for meetings; minutes of meetings; and student reports.

You won't find those topics listed under headings such as "How to Write a Cover Letter," "Model Letter for Requesting Payment," or "Standard Outline for a Progress Report." Engineered writing is a flexible approach that covers any kind of nonfiction writing, from a five-minute speech to a five-volume research report. It is not intended to help writers of fiction.

The editing techniques presented in the original book on engineered writing have been expanded to cover editing procedures for both writers and supervisory editors; to include more techniques for clarity, brevity, and simplicity at the sentence and paragraph levels; to help with correct English and current usage; and to encourage a readable personal style.

Parts IV and V of this book are new. They discuss problems and requirements arising from changing times, such as shifts in usage, the use of computers in writing, international business writing, "sexism" in language, ethnic English, and the use of English as a second language.

What has not changed since its introduction in *Engineered Report Writing* is the total system for designing written documents, for it is as serviceable now as it was 20 years ago. Part II is a fresh statement of the step-by-step procedures, greatly expanded and complete with examples and solutions to problems, political and otherwise, that writers encounter in the working world.

Nor has the concept of *engineering* the written product changed. The term "engineered" still implies well-designed and well-executed projects—in this case, writing projects that are soundly based on considerations of practical use, efficiency, economics, timing, saleability, and esthetics.

M. J. M.

Acknowledgments

Many people contributed directly or indirectly to this update of *Engineered Report Writing*—too many to name them all. Discussions over the years with research and operations people around the world have guided us in emphasizing key points throughout this book. They helped us keep focusing on practical on-the-job problems.

We particularly wish to thank the friends and colleagues mentioned below.

Dr. Elizabeth Tebeaux, Department of English, Texas A&M University, and Dr. George Binder, engineering consultant, Houston, each gave the manuscript a critical review.

Dr. Robert L. Bates, Ohio State University, provided examples of modifier trains. Alberto Noriega, engineering and economics consultant, Houston; John A. Stoddart, the World Bank, Washington, D.C.; Dr. Susan Stoddart, American Institutes for Research, Washington, D.C.; and John Schlosser, financial and marketing consultant, Seattle, made many helpful observations on international business writing. Beatrice Levin, author and college English teacher, lent her expertise in a review of grammar sections.

Eduardo G. Garza, Houston, created the cartoons and drafted the other illustrations. Grace Hay-Roe entered large blocks of the first draft into the Tandy Model 4 computer on which this book was compiled, using the Tesler Software Corporation's splendid word-processing program, ALLWRITE. Our editor, Kathryne Pile, was the epitome of tact and patience; she and the PennWell staff made numerous improvements in text and layout.

Finally, we want to thank the men and women who have participated in our executive seminars and writing workshops over the past quarter-century, and the companies that sponsored those sessions. They have made their experience ours—their problems and goals, their failures and successes. And they have fully justified our belief that essentially all professionals *can* communicate with craftsmanship. In a sense, this book is a reflection of what we have learned from them.

J. M.
H. H-R.

Part I

An Analysis of Our Writing
Problems

Does This Sound Familiar?

This manual represents an organization and synthesis of information gained through the experience of the authors in the development of techniques for the achievement of clarity and efficiency in the production of written materials and is presented for the utilization of scientific, technical, student, and business writers in the expeditious preparation of reports, letters, memoranda, and speeches. It is hoped that this compilation will not only provide a means of facilitating ease of preparation for the user but will also result in a considerable improvement in the quality of writing being produced for both internal and external publication, along with a parallel improvement in the quality of oral presentations. [109 words]

Here Is What We Mean:

We hope this book will make it easy for you to write first-class reports, letters, memos, and speeches. [19 words]

Part I _____
An Analysis of Our Writing Problems

A pressing question: Why is so much technical, scientific, and business writing like the first paragraph on the opposite page?

It looks familiar, doesn't it? We all know the damage such tiresome prose does. At worst it goes unread; at best it strains the reader's attention. Small wonder it so often fails to get action.

Further, if we multiply the number of words in that paragraph by 10 (about the length of the average technical memo, hundreds of thousands of which are written daily), we begin to get some idea of the enormous amount of time and money invested in communications and the waste that may be involved.

Why *do* we write that way? Why do we fear that departing from that stuffy style might invite criticism from professionally important people?

Let's take a look at the influences that have molded our writing attitudes and habits—in order to make some changes.

FACTORS THAT INFLUENCE
OUR ON-THE-JOB WRITING

Of the many factors that create problems for on-the-job writers, we find nine that merit discussion in this section.

Early Training

As high school students, we practiced a kind of writing that taught basic lessons about the language. It was never intended for practical, on-the-job communication.

The good English teacher encouraged us—with excellent reason—to acquire and use new, big, and unusual words. We were pressed to experiment with complex word arrangements. In this way the teacher helped us build a vocabulary and develop a style for creative writing. We learned how to create suspense, to build up to a dramatic climax, to stretch a half-baked idea into a 500-word essay.

The competent English teacher cautioned us, however, that suspenseful writing is not suitable for all purposes, that brevity is usually a virtue, and that we should exercise judgment in using our newly developed vocabulary. We should have practiced writing with the simplicity and economy that distinguish a craftsman from a beginner. (To appreciate why we have stayed with the standard word *craftsman* rather than use the neologism *craftsperson,* refer to Part V, p. 237.) But for one reason or another, many of us never acquired that skill. The need for it came as a jolt—on the job.

The writing habits so laboriously acquired in high school were reinforced in college. There we chose a field and learned technical vocabularies that only our peers could understand. Moreover, we learned to organize information in a backward, self-serving fashion that elevated our *labor* above the useful *product* of that labor.

By the time we started our careers, we had unknowingly acquired writing habits that actually block communication. We might not know a verb from a gerund, but we have become experts in converting the forceful form into the weaker one. We have learned to embroider the language and pad our writing with important-sounding words we would not dream of using in conversation. Polysyllabic nouns have become the normal form of expression in technical writing, and pomposity now characterizes the accepted professional sound.

At least that is the way it has been for many years. But fortunately for future on-the-job writers, **times are changing.** Many college professors are encouraging students to write reports that not only provide proof of academic progress but that also develop the student's ability to deliver a clear and concise message. Educational institutions now offer some excellent writing courses. If you are a student, take advantage of all the help a special course— or a special professor—can give you. After a few years off the campus, you will see that recognition often eludes skilled professionals who cannot communicate important concepts, while less competent persons are recognized because they can *sell* their modest ideas.

If you have been out of university several years and find your writing fails to mirror the careful work you do, help is available. You can improve your writing skills dramatically, regardless of your background, experience, or environment. This book's "engineering" approach to writing is simple, easy to apply, and practical.

Further help may be available within your organization. There is a widespread push for better writing in most large institutions, and many offer on-the-job training. Henry Ford II expressed the attitude typical of executives when he was asked what he looked for in young executives:

> For starters I'd list honesty, candor, good judgment, intelligence, imagination, and *the ability to write clear, concise memos* [italics ours] . . . I'm particularly pleased to see someone make a point in a simple, forceful manner . . . I have an aversion to pretentiousness and long-windedness.*

If you are never satisfied with what you write, if you must rewrite frequently, or if your efficiency rating is low because of inadequate communication skills, do not wait to be selected for a company training course. Volunteer—fast.

If you *are* selected for a writing workshop, do not conclude that you are being picked on because you are the worst in your group. At the start of a workshop, an engineering manager in an outstanding people-oriented company explained his nominations like this:

> "Yes, you've been singled out, but not in the way you might fear. Our company sees you as its future leaders. For that reason we want to give you all the skills you will need to do your job effectively and to advance quickly. Writing well is one such skill."

Sacred Tradition

"It has always been done that way."

What a wretched excuse for boring, hard-to-understand reports and professional papers! We hear that excuse even in research organizations, where open-mindedness is in the tradition of good researchers—where off with the old and on with the new-and-improved is the name of the game. We hear it from the timid and from the unimaginative. We hear it too often.

Traditional scientific reports follow the King of Hearts' advice to the rabbit in *Alice in Wonderland:* "Begin at the beginning, and go on until you come to the end: then stop." With this kind of organization, the reader suffers as the author did, through all the

---

*Milton Rockmore, "Management Shopping List," *American Way* (in-flight magazine of American Airlines), October 1979, p. 51.

phases of a project from statement of a problem through work history (including blind alleys and negative results) to the delayed solution.

Furthermore, that delay usually makes it impossible to extract from the terminal summary a single sentence stating the NEWS so clearly that it could stand alone sensibly at the top of Page 1. Writers are afraid the message will not be accepted if the groundwork is not properly laid; therefore, they rely on the persuasive power of a buildup to make any conclusion other than their own unthinkable. After such a buildup, writing a plainer statement would be talking down to readers.

We don't want to talk **down,** but we don't want to talk **up,** either. We need to talk **to.**

Granted, (1) the tone of a scientific report or professional paper for a scholarly journal is more formal than that of a newsletter or an internal memo; (2) the subject matter is specialized and difficult; and (3) the language of specialists can be used to communicate with specialists. But none of this says the document has to "impress" (read that as *de*press) by virtue of suspenseful organization and sentences crammed with difficult concepts. Scientists have the same needs as other readers. Their time is as valuable as anybody else's. They, too, benefit from clear, concise writing that elevates main points above detail. And scientific managers and executives simply have no time for wading through long, suspenseful reports.

Scientific reports *can* be crisp and direct, as demonstrated by outstanding scientists. An excellent example is a paper in the *Journal of Comparative Physiology.** As excerpts from the opening summary show, we do not even need to understand the subject of the paper to recognize that knowledgeable readers are getting the NEWS (conclusions) immediately:

2. Under specified conditions . . ., the output frequency of such networks depends weakly on the number of more-or-less similar generators in the network.

3. Given certain assumptions about . . ., networks whose function is to encode information improve their overall

---

*Michael J. Murray, "Systems of Mutually-triggering Event Generators: Basic Properties and Functions in Information Transmission and Rhythm Generation," *Journal of Comparative Physiology,* 117, 63–98 (1977), © Springer-Verlag, 1977.

signal-to-noise ratios with increasing numbers of generators. . . .

6. Mutual triggering between different spike-initiating zones
   in the lateral-line afferents of *Xenopus laevis* cannot at
   present be shown to improve signal-to-noise ratio, sensi-
   tivity, or robustness—its function remains unknown.

Direct, specific writing is good writing, even if it is not traditional.

"What the Boss Wants"

Many of us feel a twinge of anxiety when we start writing for
the approval of supervisors. What do they want? A valid question;
a very pertinent one. We should certainly analyze the supervisor's
interests; that is an important part of the procedure described in
Part II.

It is a mistake, however, to assume that all supervisors demonstrate, in their own writing, what they want. Many are like a
manager in one of our seminars, who deplored the fact that he was
stuck with his poor best. "I'm not a good writer myself," he said.
"Nor am I an editor. I can't tell them what to do. But I know bad
writing when I see it, and I don't want it."

A writer in a large company was once quizzed about his style.
"Do you like your writing?" he was asked. "No," the author
replied. "It's too stuffy. But that's the way The Boss wants it."
In a follow-up interview The Boss admitted, "It's deplorable writing. But we have to write the way The Big Boss likes it." When
The Big Boss was questioned, it became clear that writing like the
boss is not necessarily the best way to make points. "I don't know
why our people write in this stilted manner," he said.

As a matter of fact, it is a mistake to copy anybody's writing,
good or bad. Mimicry produces a caricature of the original style
and a mongrel version of the writer's own.

What *do* supervisors want? As a rule they have limited requirements: a document must be technically accurate and complete; it
must be clear and concise; and the grammar and usage must be
acceptable. All must reflect favorably on the supervisor, his group
or department, and the organization as a whole.

However, honesty compels us to recognize that there are exceptions to the rule. Some supervisors really do, in effect, demand
ghost-writing, for they insist upon hearing their own voices in everybody's writing. Perhaps they were born to be writers or editors

and somehow got into the wrong field. No matter what you offer, they will make changes. Indeed, if a piece of their editing has been out of their hands for a while, they are quite capable of deleting their *own* editorial "improvements" from a retyped draft.*

If you find yourself obliged to serve as a ghostwriter, you have our sympathy; but you need more than that. Ghost-writing is a special profession requiring talent, training, and suppressed pride of authorship. Probably the best solution to this problem is not to lose any sleep over arbitrary changes plainly based on personal preference, unless they render the text inaccurate or ungrammatical. Then a conference is in order.

We end this discussion with a word of **caution:**

Before charging out to show somebody how your prose has been mutilated by a supervisor, study those changes. There may be good reason for them. Do they add information that you should have thought of? Do they improve organization? Pull split topics together? Correct grammatical errors? Shorten or simplify? Ease awkward constructions? Strengthen verbs? Correct your modification? Take into account a problem of office politics, corporate policy, or diplomacy? Keep an open mind as you study changes. If you still disagree with them, then ask for a discussion.

The "Peon Complex"

An uncomfortable feeling shared by many younger professionals (and more than a few older ones) is what we call the *peon complex*. The term was invented by self-styled "peons" in our workshops. They say—

I'm not supposed to make recommendations; I can't tell the manager what to do. I'm only a peon around here.

I can't make the decisions. That's what the boss gets paid for.

I've been here only a few months; they wouldn't be interested in my opinion.

I'm just laying out the facts. That's all peons like me are supposed to do.

---

*Does office gossip cast you in the role of such a supervisor? If so, you may be interested in "Some Techniques for Supervisory Editing" (p. 132).

A junior insurance analyst in a writing workshop wrote an 11-page report to her boss on various types of insurance—descriptions, terms, prices, options, advantages, and disadvantages. It was an extremely comprehensive memo on a thorough investigation. But she made no recommendations and offered no conclusions. At the end of the report she signed off with, "If you need further information, please feel free to call me." The manager probably did not need the invitation, but he did need further information.

Nobody in the workshop could convince the analyst that a conclusion or recommendation was in order until her manager, unobtrusively monitoring the class from the back of the room, spoke up: "Janet, I passed this job to you for two reasons. First, you're our expert in the field and I trust your judgment. Second, that's your job, and I don't have time to do it. Are you telling me now, after checking the facts as thoroughly as your memo suggests, that you still can't give me any help?"

To encourage the peons, one division office of a major corporation found it necessary to issue a directive entitled "Finished Staff Assignments." It stated, in part:

A staff assignment is not finished until the problem has been studied and a solution presented in such form that all your supervisor need do is indicate his approval or disapproval of the recommended action. . . .

It is your job to advise management what to do, not to ask them what to do. Supervisors need answers, not questions. . . .

The concept of finished staff assignments does not preclude rough drafts . . . but they must not be used as an excuse for shifting the burden of *formulating action* [italics ours] to the next higher level. Nor should the draft be given to your supervisor as an editing task for him. . . .

This approach may mean more work for you, but it gives more freedom to your supervisor. This is as it should be.

Insecurity

Some writers feel naked and insecure with brevity and simplicity. They need to justify their professional existence by the length, heft, and jawbreaker language of their writings.

One outstanding engineer had this problem to an exasperating degree. When asked for a memo on a given topic, he would tell the secretary to type the memo on legal-size paper, which he considered more impressive, specifying "one-and-a-half" spacing to make the memo seem longer. Then he would go to the files, dig out his old memos on related topics, photocopy them, and attach them to the requested memo to make a bulkier package.

Needless to say, the result was depressing rather than impressive, and he was finally told to stop mistaking quantity for quality.

This problem is not limited to beginners. An insecure supervisor, upon receiving a draft "only" 10 pages long, once counseled his subordinates this way: "It took three people, three months, and $50,000 to complete this project. I want your report to reflect that investment. Double its length."

Of course a practice of this sort needlessly squanders credibility, not to mention manpower and funds. If work is good, it needs no embellishment; if it is worthless, no amount of padding will fool an intelligent reader. Perhaps the only possible excuse for deliberate padding is to hide some really bad work in such intimidating bulk that nobody will ever read far enough to find it.

If you do good work, feature it; keep details lean, tidy, and as brief as you can. You want your document to receive prompt attention at the highest levels possible. Consider the busy manager in the drawing. Assuming the subjects are of equal urgency, which document do you suppose she will read, and which will she pass on to an unfortunate subordinate? Which would *you* pick up first?

WHICH REPORT WILL SHE READ FIRST?

Too Many Cooks in the Kitchen

In large institutions, documents frequently reflect the contributions and editorial changes of not one, but three or four supervisors. Sometimes the resulting products turn out like a soup that was seasoned by too many cooks. Authors become discouraged and begin turning in carelessly written material, thinking, "What does it matter? It's all going to be changed anyhow."

It does matter, of course. Their names or initials go on such documents, and it is those authors—not their supervisors—who will be credited with poor quality. Moreover, file copies of badly written documents can haunt an author for a long time.

Instead of becoming cynical about multiple editorial changes, authors would do well to eliminate the need for them. The Pre-Writing Conference, described in Part II, is a tested way of doing this.

The Editorial Decree

Many companies and institutions have professional editors; if all else fails, we might blame our writing difficulties on the in-house editor. Editors are the author's ghostwriters. They rewrite or reorganize impenetrable passages; clarify foggy sentences; correct grammar, punctuation, and spelling errors; meet deadlines; and insist on reasonable standards for correctness, clarity, and uniformity—all for the glory of the author with none for themselves.

But even editors can sometimes be part of the problem. Editors compose rules and instructions. Like any other writer, they labor over the creation; consequently, they sometimes fall in love with it and forget that the creation was intended to serve writers, not vice versa. If editorial rules restrict freedom to organize to suit the author's purposes and the reader's interests, we may expect resistance from conscientious writers.

On the other hand, many seemingly arbitrary editorial rules are established for good reason. They can provide uniformity within documents written by several authors or edited within a group of several editors. And they may be necessary for efficiency and economy in production. Such rules, of course, deserve acceptance.

Arbitrary Guidelines

Until recently most company manuals and professional journal guidelines were rigid affairs, dictating how documents had to be organized regardless of subject, purpose, readership, or information available. Prescribed outlines left nothing to the writer's intelligence or common sense. Such guidelines, especially those published by prestigious journals, have deeply influenced our attitudes toward writing. They have also been responsible for a great deal of ineffective communication.

For example, conformance with an outline like that in Fig. 1 means that some writers must pad a lot. Not every article needs the mandatory "Introduction," or "Scope of Work," or "Materials and Equipment." Thus, an author would have to dredge up or invent filler material. This padding reduces readability, increases length, and runs up journal page costs, all of which are presumably of great interest to journal editors.

It is safe to say that most editors try to encourage simplicity and brevity in the articles they receive for publication. A random sampling of organizations whose editors or editor-members are concerned about clear and concise writing would include the Society for Technical Communication; Westinghouse Electric; IBM; the American Gas Association; the Association of Earth Science Editors; the U.S. Navy; Texas state agencies; the Cana-

I. ABSTRACT

II. INTRODUCTION

   A. Purpose of Study

   B. Scope of Work

   C. Previous Work

   D. Acknowledgments

III. MATERIALS AND EQUIPMENT

IV. FIELD METHODS

V. LABORATORY PROCEDURES

VI. FIELD DATA

VII. LABORATORY DATA

VIII. DISCUSSION

IX. CONCLUSIONS

X. RECOMMENDATIONS

Figure 1   What's wrong with this outline? (For an answer, turn to p. 50.)

dian Law Information Council; the Society of Petroleum Engineers; and the American Association of Petroleum Geologists.

Nevertheless, there are still some unduly restrictive guidelines and company manuals around, and you may have to submit to their rules to get published. But you can improvise a bit in a fashion that will improve readability, cut article length, and save money for publications as well. An illustration of this technique is presented on p. 50 under "Having to Follow Unsuitable Guidelines."

You can sometimes judge a style manual by its rules on current usage. If it includes such rules as those below, you might well wonder whether it is authoritative and up-to-date.

Never end a sentence with a preposition.

Of course you can. A preposition is often the best word to end a sentence <u>with</u>.

Never split an infinitive.

Nonsense. <u>To</u> deliberately <u>split</u> an infinitive is sometimes effective, even necessary, for clarity. Be careful, though; when too many words intrude between the <u>to</u> and the verb, the sentence can become very awkward.

Never start a sentence with *and* or *but*.

<u>And</u> that's another piece of nonsense.

Never separate a verb from its auxiliary with a modifier.

Who says we <u>can</u> never <u>separate</u> them?

Go ahead and ignore outdated rules that inhibit plain contemporary writing. But be sure they are indeed outdated, lest you ignore something that makes your writing correct and tasteful. If in doubt, look it up. Go to the library and check out a book on contemporary English usage. (See the Selected References on page 255 for some good ones.)

Language Deficiencies

Nothing is more likely to turn prose stiff than a writer's knowledge that his language skills are deficient.

We might say that educated adults can write and speak good English without knowing how or why; therefore, they need not worry about usage and grammar. In some cases this is true. Many learned English in its best form at home, much as they learned how to dress. They associate with cultivated friends; they are devoted to good literature and have time to read it; they hear standard English on network news broadcasts (although the effect is blunted by exposure to the substandard English of many commercials); and good English is spoken daily in their offices. If they are lucky, they write it and speak it well because they are intuitive about what is good and what is bad.

The intuitive writer does not need a grammar book or a set of rules. Nevertheless, we must all evaluate our gifts and talents with objectivity. Are you intuitive? Try the following test. If you are thus gifted, a small voice inside you will say which sentences should be revised—where ''something'' is wrong.

Test 1
CHECK YOUR INTUITION

How many of the following sentences are grammatically faulty? Mark the correct ones with a **C.** Mark the faulty ones with an **F.** Then check your answers in Appendix A (p. 259).

(1) This hospital expects an emergency patient to recognize that they don't have all the answers.

(2) The printer was connected to the computer, but it was found to be defective.

(3) Heating costs were reduced by thicker insulation and lowering the thermostat.

(4) The two solutions were injected into the model and the liquid withdrawn.

(5) The process has the capacity of atmospheric pollution reduction.

(6) By enlarging this annular passage, the controlling valve can rotate relative to its housing.

(7) A spokesman said Deep South Airlines is as safe, if not safer, than it has ever been.

(8) Reduction in death and serious injury rates occur as a result of this change.

(9) If a high energy had existed it would tend to have broken up the solids.

(10) This modification is important in processing, and we have designed a parts kit, also.

(11) In fact, a person has hurt themself on this equipment in the last 24 hours.

(12) This device is used to reduce the scale and in reading the values.

If you are not 100 percent intuitive, be assured that you have plenty of company. Here are four examples from real life:

CASE HISTORY 1

Dr. A (PhD in chemistry) learned to use acceptable English in a painfully careful way. But he viewed language as a mysterious subject he never mastered and never enjoyed. Any communications situation paralyzed his power of expression. He could not deliver his message, either in writing or in oral presentations. His prose was stiff, stuffy, too complicated to be understood.

The division manager said Dr. A was an excellent researcher. However, somebody else had to sell his ideas in meetings and reports, and he seldom received credit or promotions. Dr. A appeared to be withdrawn and unsociable. If you got to know him, though, he was simply a nice person with a bitterly disappointing career.

CASE HISTORY 2

Biology Professor B grew up in a small town in Saskatchewan where cultivated English was rarely heard. He came to the U.S., and after getting Bachelor's and Master's degrees at an obscure southwestern college, he taught for a number of years at three junior colleges. Up to this point, his shaky command of English ("them there samples"; "if I'da knowed that, I woulda went") had never been a handicap. Then Professor B took a leave of absence and entered a top-notch state university to work on a PhD. As a new student, he was required to take a routine exam in English, along with all the entering freshmen. To his chagrin, he flunked the test and was obliged to sign up for remedial English.

Mortified, he worked hard to overcome 35 years of bad habits. Gradually his English improved. He earned a PhD and until his recent retirement, headed the Biological Sciences Department in a midwestern university.

CASE HISTORY 3

When geophysicist C was hired by an oil company, she was considered technically one of the department's most promising recruits (4.0 average, *summa cum laude* graduate). Then it was discovered that the campus slang noticed in her interviews had carried over into her job performance. Her letters were ungrammatical and her oral presentations were almost unintelligible.

Her supervisor was not impressed when Ms. C explained that she was "lucky" to have had a "contemporary-type English prof, y'know." This professor "y'know, he was like, well, very liberal—like he didn't really go for . . . like . . . *who* and *whom* distinctions and like all that. He sorta believes that old-fashioned grammarians are like, well, tyrants."

But Ms. C really is a bright young woman. She signed up for a technical writing course and then set herself a program in English grammar. Her supervisor has not been disappointed.

CASE HISTORY 4

Engineer D's English was substandard in both speaking and writing. He said he had faked out freshman English in college and had learned all he knew about language long after he came off the range and brushed the burrs from his pants. Mr. D, overheard in conversation: "Has it came by your office yet?" "I throwed mine in the wastebasket." "I ast Allan did he have his yet."

Engineer D's exceptional productivity saved his job, but he was a source of embarrassment to his company (and to himself) on many occasions. He now feels that his language deficiencies have greatly limited his career.

It would be unrealistic to pretend that Dr. A, Professor B, Ms. C, and Mr. D are rarities among educated professionals. They are not; and they deserve help, not flattery. Any writer with recognized language deficiencies should follow Ms. C's example. Part III of this book is designed to help writers, whether they have basic deficiencies to overcome or merely want to refresh their memories. The books listed in the Selected References can be enlightening; the grammar books recommended there will be especially helpful to writers who want to understand our language as well as follow rules.

HOW TO SOLVE OUR WRITING PROBLEMS

So many factors influence our attitudes and habits in writing that a simple solution appears to be out of reach. It isn't.

By recognizing and analyzing the problems behind ineffective communication, any professional can develop an enthusiastic new attitude toward what has been an unwelcome chore: on-the-job writing. This change makes it possible—perhaps even easy—to improve.

The question, then, is how to upgrade the writing. What follows in this book is a total "how-to" system for achieving this goal—and for doing the job with a big cut in communications costs. The system includes:

- A conversational method of organizing ideas, based on the familiar pattern of normal dialog that we all practice. This logical approach helps writers to get to the point at all levels of a document, reducing length by as much as 75 percent (in lengthy texts) without sacrificing desirable content.

- Efficient review and approval procedures that (1) eliminate needless recycling and rewriting of drafts, with a corresponding boost in writer morale, and (2) reduce management review and approval time (by up to 95% in large corporations). This saving of time and effort improves morale among both managers and writers.

■ Editing techniques for both authors and supervisors, based on courtesy, common sense, and an adequate knowledge of grammar and current usage.

Let's conclude this section in the terms of our opening ''Memo to All Professionals'': We hope this manual will help you write first-class reports, letters, memos, and speeches with ease and personal satisfaction.

Part II

How to Organize the Message:
The Murray System

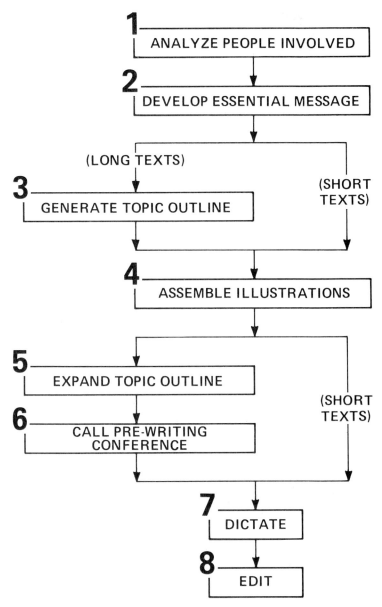

Figure 2 The Engineered Writing System

Part II
How to Organize the Message:
The Murray System

> Of all the sacred cows which graze in the field of medical writing, perhaps the most sacred is The Outline. . . . To work around a rigid skeleton, as an inviolable framework, is stultifying.
>
> —Lester S. King, MD*

A NEW APPROACH
TO PRODUCT DESIGN

Although the idea of engineering a product is familiar to technical writers, the concept of engineering the design of a document (Fig. 2) appears to be a novel one. There is, however, little difference in the approach to planning: whether you are designing a hospital, a computer program, an offshore drilling platform, or a scientific report, you must consider practical use, efficiency, economics, timing, saleability, and esthetics.

In common writing practice, these factors have been disregarded in favor of tradition. On-the-job writers have typically followed the suspense-filled "Biblical" approach—going from Genesis to Revelations—illustrated in Fig. 3. This triangle represents the structure of a formal report, but the same suspenseful pattern appears in letters, memos, and speeches.

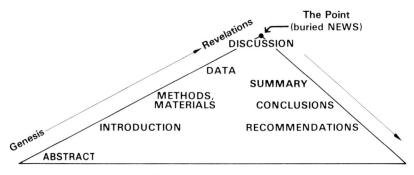

Figure 3 The Biblical or suspense format.

*Lester S. King, MD, and Charles G. Roland, MD, *Scientific Writing,* AMA Publications, 1968, p. 64.

With a ready-made model like this, organization requires no judgment, no common sense, and very little intelligence. Anyone can do it. If readers do not understand, then the subject must be over their heads. The problem is theirs.

There *are* occasions to choose the Biblical format. A special problem (pp. 43–46) may necessitate some degree of suspense. Mystery stories, novels, and most histories certainly cannot "get to the point" on Page 1. Moreover, some professional journals still insist upon the suspense pattern, and we have to manage with it if we want to be published in those journals.

But in scientific, technical, and business writing, the indiscriminate use of the suspense format for organizing ideas has been wasteful and ineffective. Documents cost far more time, money, and effort than they need to; they fail to mirror the careful work that is the subject of the writing; and after several laborious revisions, they are thrust on a reader who must embark on a treasure hunt for the useful ideas, armed only with a yellow Hi-Liter® pen.

To find the main point of a document organized in the traditional manner of Fig. 3, we may be forced to search through a series of off-the-shelf units:

Abstract: Often misused to describe the shape of a report or article (like a narrative table of contents) rather than its essence. Called a **descriptive abstract** (Fig. 4A), it will feature passive verbs such as "are discussed," "is presented," "are reviewed," and "is described." An **informative abstract** (Fig. 4B), intended primarily for technical readers who may go on to read the body of the document, gives the essence of the entire text. An **executive summary** (Fig. 4C) focuses on information of importance to managers, who probably will not read the entire document.

NOT THIS —

┌─────────── **A. DESCRIPTIVE ABSTRACT** ───────────┐

Following an analysis of the need for better written communications, an eight-step method for organizing any kind of nonfiction written or oral presentation is described by this book. Many examples are included. Also presented are numerous techniques for improved writing at the sentence and paragraph levels, with discussions of the problems of communication between North American scientists and business people and their correspondents overseas. Finally, problems peculiar to the last quarter of the 20th century are reviewed.

BUT THIS —

┌──────────── **B. INFORMATIVE ABSTRACT** ────────────┐

Based on the pattern of good dialog, the Murray System enables writers to deliver their message clearly and concisely with less effort, but with maximum impact on the chosen audience. The method has from three to eight steps, depending on what is being written:

(1) Analyze people involved   (5) Expand topic outline

(2) Develop essential message   (6) Call pre-writing conference

(3) Generate topic outline   (7) Dictate complete text

(4) Assemble illustrations   (8) Edit and polish

To improve one's writing at the sentence level, the most important technique is to use verbs effectively, preferring active verbs. At the paragraph level, help the reader with summarizing topic sentences and transition words and phrases.

└──┘

OR THIS —

┌──────────── **C. EXECUTIVE SUMMARY** ────────────┐

The eight-step Murray System for engineering the design of reports, letters, memos, and speeches enables users to improve readability while cutting text length, speeding up the writing and approval of in-house documents, and cutting costs. Organizing techniques for on-the-job writers and editing techniques for supervisors make written communications more efficient for both writer and reader.

Productivity increases are achieved through:

- A conversational method of organizing ideas, based on the familiar pattern of normal dialog that we all practice. This logical approach helps writers to get to the point at all levels of a document, reducing length by as much as 75 percent (in lengthy texts) without sacrificing desirable content.

- Efficient review and approval procedures that (1) eliminate needless recycling and rewriting of drafts, with a corresponding boost in writer morale, and (2) reduce management review and approval time (by up to 95% in large corporations).

- Editing techniques for both authors and supervisors, based on courtesy, common sense, and an adequate knowledge of grammar and current usage.

└──┘

Figure 4   Three types of abstract for this book

Introduction: May include a lengthy description of the "Problem," even if the *reader* originally submitted the problem; "Objectives," although they may already have been attained with spectacular success by the time the report appears; "Background" or "History of the Project," whether useful to the reader or not; and "Previous Work," including failed attempts and dead ends.

Methods/Procedures; Equipment/Materials: Although the reader commonly wants to know what the writer found out, and only secondarily *how* he did it, these sections of a report often do their part in delaying the real NEWS.

Results/Discussion/Data: A detailed description of work done and data generated, the interpretation of the data, etc. Look at Fig. 3 now. You have labored up the slope of the triangle to find the NEWS; it is probably buried here somewhere. If you are lucky, it will be plainly stated. At worst, it will only be implied.

Summary, Conclusions, Recommendations: At last we have come to the real NEWS—or have we? After sprinkling crumbs of information all along the way, is the author likely now to come

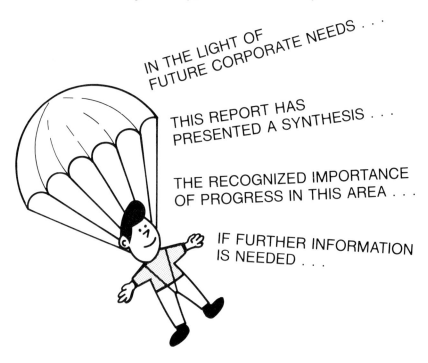

IN THE LIGHT OF FUTURE CORPORATE NEEDS . . .

THIS REPORT HAS PRESENTED A SYNTHESIS . . .

THE RECOGNIZED IMPORTANCE OF PROGRESS IN THIS AREA . . .

IF FURTHER INFORMATION IS NEEDED . . .

right out with a plain and simple statement of his NEWS that could be moved to Page 1 to stand alone meaningfully? Almost never.

And a "Parachute": Reluctant to stop abruptly, the timid writer will add a parachute sentence (or paragraph) to let the reader down gently.

Books have been written to help readers get efficiently through texts organized like this. A good example is *How to Read a Book.** It has sections entitled "Finding the Key Sentences," "Finding the Arguments," and even "Finding the Solutions" in order, as the authors point out, *"to determine the message"* [italics ours]. That we actually need such help is a sad commentary on the state of written communication.

THE ENGINEERED
TEXT IS DIFFERENT

Well-engineered documents turn that traditional outline upside down, highlighting the key sentences, the solutions, the message—the writer's real NEWS. As suggested in Fig. 5, main points are elevated above detail, not only at the beginning of the document, but also at the beginning of each of its major parts.

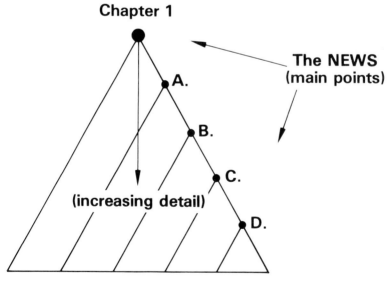

Figure 5   A reader-oriented approach to organization.

*Mortimer J. Adler and Charles Van Doren, *How to Read a Book,* (New York: Simon & Schuster, 1972).

Whether writing a letter of a few paragraphs or a formal report of many chapters, we begin at the peak of the triangle (Fig. 5); we do not force the reader into a long climb to reach the NEWS.

The benefits of this direct approach are worth reviewing:

- The method reduces text length by as much as 75 percent (in long reports) without sacrificing desirable content.
- It eliminates rewriting and recycling of drafts (for an obvious boost in writer morale).
- It reduces review and approvals time up to 95 percent in large organizations; this improves morale among managers as well.
- It cuts production costs substantially.
- And it is easy to apply to all kinds of nonfiction writing.

The Murray System involves from three to eight steps, shown in Fig. 2 and explained in the sections that follow. You will find examples of how the procedure is applied in Steps 2, 3, and 5, with a complete example for a long report in Appendix B.

(1) WHAT is your NEWS?
    (a) What do you (the author) WANT?
    Or
    (b) What do you OFFER?

(2) What are READERS' INTERESTS?

	Name and Job Title	Chief Interests
(a) To [addressee(s)]:		
(b) Through [persons on addressee's staff]:		
(c) From [person who signs, if not the author]:		
(d) Through [persons who will review/approve before document is signed]:		
(e) Copies to [recipients other than those listed above]:		

(3) Do you foresee any SPECIAL PROBLEMS IN WRITING?

If so, how will you handle them?

Figure 6   Step 1—analysis of people involved.

Step 1
Analyze the People Involved

Good communication starts here—with the question WHO?

Few of us got through high school without having heard this firm pronouncement from an English teacher: **consider your reader.**

Most on-the-job writers do "consider" readers, to the extent that they either put a name at the top of the first page or ask a secretary for a distribution list. Beyond that, they expect that anyone interested in reading what they write would be interested in reading *whatever* they write.

That cursory sort of consideration almost guarantees that the writer will lose sight of readers as individuals with job responsibilities that create specific interests in what a report, letter, memo, or article has to say. Losing sight of people means losing sight of practicalities.

The practical purpose behind every document will involve at least one main point that responds to the reader's interest. But identifying and getting to that point are not the easiest parts of communication. At the outset of a writing project, the author's head may be crammed full of cherished information, all of it so interesting that the author could start almost anywhere: with what he thought about, what he did, why he did it, or how he did it. To the author it all seems important and informative. But it is not necessarily what the reader wants or needs to know.

The Analysis of People (Fig. 6) is a systematic method of determining the main point that every reader would like the writer to reach. It defines (1) the writer's objectives and (2) what the other people involved (principally as readers) want to know. The analysis thus gives a concise statement of the NEWS (main point) that serves all readers, sets up the organization of the entire text, and makes the writing a relatively easy task.

The readers considered include not only those to whom a document is addressed, but also those who approve it or sign it and those who will receive copies. They too have requirements that affect the form and content of a document.

An important aspect of the analysis is recognizing special problems that affect (1) the writer's attitude (as well as the readers') toward a subject, (2) the writer's ability to make decisions regarding content and emphasis, and (3) the quality of the finished document. These special problems are discussed later in this chapter.

29

Most of us prefer to bypass this step in communication because we tend to view writing as an intrusion into more pressing or profitable pursuits. Taking time to consider readers in detail seems like imposing an added hindrance to getting the job done.

But this early investment of time pays powerful dividends: in the long run, it saves more time than it costs; it ensures appropriate organization for the people involved, writer as well as readers; and once the analysis is complete, the rest of the writing job is an easy, downhill glide all the way.

HOW TO MAKE
THE ANALYSIS

Complete the form in Fig. 6 to determine your NEWS. Until you are practiced in using the system, don't try to do this analysis in your head. Always write it out for documents long enough to require an outline.

The content of Fig. 6 is based on two practical facts. First, every reader wants to know one of two things:

(a) What do you want FROM me?

Or

(b) What do you have FOR me?

Second, all writers have exactly corresponding objectives:

(a) They want something FROM the reader

Or

(b) They have something FOR the reader.

If common sense prevails, either point above, or both, will define the NEWS of the document. About 98 percent of the time, this is what your document should begin with.

Let us consider the questions in Fig. 6.

(1) WHAT is your NEWS?

(a) What do you WANT?

Realistic answers to this question might be worded like those that follow (with formal phrasing in parentheses):

—I want $258,500. (We request approval for an expenditure of . . . .)

—I want to improve our recruiting practices. (The Personnel Department recommends that the following changes in our recruiting practices be implemented immediately: . . . .)

—I want Accounting to give me statistics on . . . . (Would you please furnish . . . .)

—I want to drill an exploratory well. (Southern District Exploration proposes . . . .)

It can be surprisingly hard to pin down the real objective. When writers are asked what they want, they are apt to launch into a long discussion beginning with, "I want to tell them **about** . . . ." Like the engineer who wanted to "tell my manager **about** the down-time on No. 3 generator," although his real point was "I want money to buy a new generator." Or like the oil-company writer who said, "I need to tell them **about** the results of this geological study," when the real message was: "It's a bonanza! Let's start drilling!" Or the systems analyst who said "I want to tell him **about** some problems with our section-plotting system," when he really wanted approval to change the system:

I request approval to change our section-plotting system to move rasterizing from the host computer program to the hardware of the computers that drive our plotters.

In general, when you want something, just ask for it straight away, and **then** justify it. Unless you have special problems (p. 37), there is no valid reason for holding the reader in suspense. Give decision-makers the NEWS up front, enabling them to study the why's and how's of the matter with understanding.

(b) Or what do you OFFER?

Is it a new piece of equipment? An instruction manual? A method or technique? An answer to a question submitted? An opinion or decision? Useful data or statistics? Analysis and interpretation? The solution to a problem? Progress in a project?

Unless "progress" actually is the subject of the document, describing the *work done* does not answer this question. For example:

An in-depth study of . . . has been made.
But what was the outcome or product of that study?

A laboratory <u>analysis</u> of . . . <u>has been completed</u>.
> Unless the reader is an impatient supervisor, who cares?

<u>We</u> <u>are</u> <u>in receipt</u> of your letter of . . . in which <u>you</u> <u>stated</u> . . . .
> We are obviously in receipt, and the reader knows what he has stated. (If he is forgetful, we can use the subject line to remind him.)

As the following case history shows, separating what we want from what we offer can be the hardest part of the Analysis of People.

An engineer working for an American company in the Middle East had a typical problem. He had to write a report to the host government after an industrial accident in which toxic fumes drifted over a large area of desert, including a village of several hundred people. Naturally concerned about those citizens, who had been evacuated to tents in the desert, the government had dispatched a high-level representative to the company's local headquarters to monitor the response to the accident.

At the time of writing, the engineer-author had already found a pipeline leak and had repaired it. His company had then moved all the villagers back to their homes.

Here is an outline of the engineer's first draft:

(1) History of the plant and pipeline construction, documenting how carefully it had been designed to prevent such accidents.

(2) Description of the accident (which had nevertheless occurred).

(3) Details of the clever detective work that had located the leak.

(4) Details of the prompt repair and cleanup.

(5) Fourteen pages later, the NEWS that the area was safe and the villagers were again in their homes.

Understandably dissatisfied with this draft, the engineer presented his problem in a writing workshop. As we reviewed his Analysis of People in class, the questions and his answers went something like this:

Q. What do you **want?**

A. I want to tell them about the pipeline layout and design because I don't want the government to think our design was poor.

Q. Okay, that's what you **want.** Now let's consider this: Do you have anything of interest to **offer** the government?

A. Yes, I want to tell them exactly how this thing happened, so it won't appear that we were negligent. It was a freak accident, you know; wouldn't happen again in a hundred years.

Q. Fine; that's more of what you **want.** What do you **offer?**

A. Well, we found the leak and fixed it—I said that in my first draft.

Q. Yes, you did. But let's look at it this way: when you told the government representative about the repair, what was the first thing you said?

A. Why, I told him the village was safe and we'd promised to take care of relocating their people, and we had done that.

Q. If that was what you had to offer the government's representative. . . .

A. Oh, now I see what you're getting at!

Once this writer had separated what he wanted from what he offered, his problems disappeared. The organization of this report was established by the questions that would follow his statement of the NEWS: How do you know it's safe? How did you fix it? How do you know it won't happen again? Etc. There was a natural and logical place for every bit of information the writer "wanted to tell them." But detail was drastically reduced while nothing of significance was lost; the second draft was three pages long, instead of fourteen. The message was strong and confident (no evident defensiveness, weaseling, or hedging), and the company looked competent and caring.

(2) What are the READERS' INTERESTS?

(a) To (addressee[s])

Now that you have identified your NEWS, you can determine your readers' interests in that point. Consider your statement of the NEWS (what you are asking for or offering). If you said that in face-to-face conversation with your readers, how would each reply? "Why should I care about that?" "What's in it for me?"

A corporate executive might ask, "What's the bottom line?" meaning "How much will it make? or cost? or save? or improve my operations?"

If the addressee is a technical person, his interest could be, "Why should I do that?" or "Why did you do that?" or "How will this help me in my work? Does it solve my problem or present me with one?"

A scientist might be asking whether the NEWS alters his theory or his understanding of a theory; whether it could enable him to improve his research or continue a study; whether it might alert him to an idea he would want to question or refute, perhaps providing a topic for a professional paper.

It may now be clearer to you why Fig. 6 prompts the writer to specify actual names and job titles of the people involved. There is a sound reason for writing these things down. They will help you focus on the technical backgrounds, personal viewpoints or idiosyncrasies, and job-related interests, and thus aid you in filling out the right-hand column of the questionnaire.

Flying Blind

Suppose a friend came to you and said, "Hey, you're sort of an expert on microcomputers. How about giving a talk to our group next weekend?"

No doubt your first response would be, "Which group is that?" As a person knowledgeable on a potentially difficult subject, before preparing a talk you would want to know: Is it a group of Sunday-school kids interested in games and programming, or small-business owners who would want to know if the time was ripe to invest in a computer for their particular business? Is it a professional group whose members would already be conversant with computers, perhaps highly knowledgeable? The content and slant of your message, even its length, would obviously depend on the makeup of the audience. Asking about your listeners would be automatic.

Yet how often do we jump into a writing job (whether speech, report, or letter) without really analyzing the intended audience? This happens much more often than most people realize, even though we've paid lip service to "considering our audience" ever since school days. Without knowing the audience, we are truly flying blind.

(b) From (who signs?)

The person who signs a document (if not the author) is also a reader. We can be certain of a few interests at the outset: technical accuracy and completeness, clarity, literate English, and usually brevity. A top manager will also be very conscious of the corporate or institutional image projected by the document.

One of the problems in big companies is that managers often pass a writing assignment down to you through one or more supervisors with no more guidance than a scribbled note in the margin of a memo or letter: "Pls. handle." The manager has a specific purpose in mind when he requests that the document be written for his signature. Most of the time that purpose is obvious, but sometimes the writer hasn't the slightest notion what it might be. In that case, ask questions. Find out. It is professionally immature to write in ignorance.

(c) Through (who approves?)

Persons who have to approve what you write are closer to your subject than anybody except you. They represent all the interests of their superiors as well as their own, and they too may have a variety of objectives. One supervisor wants a technical report to create a climate for future service work. Another wants to protect against future criticism (a situation common enough to merit a familiar three-letter abbreviation). A third wants to preempt a proposal from another source.

Moreover, many supervisors foresee special problems such as conflicting personalities, technical disagreements, or antagonistic reception. (Some of these problems are discussed later in this chapter.) The chief interest of people who approve technical writing, however, is technical accuracy and completeness, as well as a standard of literacy that reflects well on their group.

(d) Who Gets Copies?

Why analyze recipients of copies? Their interests may not alter the NEWS (though copies sometimes go to top executives) or affect overall organization much. But if we consider *why* they receive copies, we may find that an additional topic or two should be discussed, or that the document should have an attachment, an appendix, or a cover letter. Many documents are returned for a

Know your audience before you go into action.

rewrite because of errrors of omission rather than commission.
Let's look at the following situations that illustrate this point:

(1) Writing a memo to the head of his biochemical research
    group, the author failed to recognize that a copy was going
    to the manager of the safety group because the writer's new
    process could present a safety hazard. In the requested re-
    write, he included a paragraph concerning that subject, plus
    an attachment outlining safety precautions.

(2) In a report to the public relations department, the author
    failed to realize there was a reason for sending a copy to the
    legal department. In his second draft he included a para-

graph describing past compliance with government regulations.

(3) Writing to her immediate supervisor, a field engineer neglected to consider the reason for sending a copy to headquarters. In her second draft she had to add an important paragraph justifying an expenditure.

Summing up: Know your readers in order to define their interests. If you don't know certain readers, ask your boss or a senior colleague—but do find out.

(3) Do you foresee any SPECIAL WRITING PROBLEMS?

The problems we discuss here have nothing to do with the work done, which may have been plagued with technical difficulties from beginning to end, and which—having been overcome—only make the results more impressive.

The kind of problems we are talking about appear at the time of writing, and they do not enhance the document; they just discomfit the author. And **they are responsible for more bad on-the-job writing than any other single factor except neglect of readers' interests.**

If a strategy for handling these special problems is not devised before the writing begins, they will inevitably be reflected in the prose. They make themselves evident in a host of organizational defects, including split topics, gaps in logic, fractured organization, discrepancies and apparent contradictions, changes in slant or viewpoint, even the appearance of weaseling or outright deception.

Of course, recognizing that problems exist will not make them all go away; some simply have to be lived with. But the mere fact of having reached that conclusion will relieve tension and strengthen your prose. You won't keep approaching the problem area, backing off, skirting it, hedging, and jumping about.

Some Common Problems

Here are a dozen special problems and ideas on handling them.

No news to report

Problem: How can I get to the point? I don't have a real point—I'm just transmitting pages and pages of nothing but statistical data.

If there really is no information to be transmitted other than
"Here it is," why try to write a letter or memo? Why not attach
a routing slip, making sure that a dated copy goes into the file?
Save a bit of time and money.

One study showed that in 1984 the average letter cost $9.00 to
$12.00 to prepare, and many writers report that they need at least
half a day to compose a cover letter or short memo. (This suggests
that they are not analyzing readers properly.)

If you should write a cover letter to accompany a lot of statistics, your Analysis of People will indicate what your readers need
to know: how the data are to be viewed or used; what conclusions
(if any) can or should be drawn from them; how accurate they are;
how they were acquired; how they have been organized; and so
forth.

> **Problem:** How can I get to the point? I'm just documenting
> something for the files—sort of a history.

If the project (the subject of the history) led to any conclusions or
insights or lessons for the future, such information could be as
valuable to future readers of the document as to present readers; it
may help them avoid reinventing the wheel. Start with that information and follow with the history. You can then, as a historian,
feel perfectly comfortable with the "Biblical" approach. History
does begin at the beginning and proceed toward the end.

> **Problem:** In this progress report, how can I get to the point
> or state any news at all? I haven't made any prog-
> ress this month.

We've never found an entirely satisfactory answer for this one.
Since progress reports are usually written by edict at specified intervals, writers often feel obliged to come up with something or
suffer unpleasant consequences.

In order of preference we would choose the following:

(1) to skip one monthly report; but if that is not permissible,

(2) to write a one-liner (not the comedians' kind). For example:
    "Collection of data continues on schedule, but it is too early
    to draw conclusions." Or "The field work is now 40%
    completed." Or "This project has been held in abeyance
    for priority work on . . . ." If that won't fly,

(3) to reach for trends or preliminary indications (weak though they may be), or perhaps to discuss plans for the coming month based on status of the project.

In any case, keep it brief.

Shaky or insufficient data

Don't let this problem make you defensive, verbose, or circuitous in your writing. Select your verbs deliberately. The auxiliary verbs *may, could,* and *should* are often helpful, as are adverbs like *probably, evidently,* and *apparently.* But don't overdo the qualifying! Don't qualify so often and so heavily that nothing remains, as one author did:

The data appear to indicate that the theoretical relationship may apply in this instance.

Be especially alert to the fact that apologetic writers tend to obfuscate, as was the case with the author who wrote this:

Despite the inadequacies of the available information, a careful speculation as to the possibilities is necessitated by the

---

A Precision Instrument

Properly used, the verb is a high-precision tool you can use for "fine-tuning" your sentences. Here, for example, is a sequence of verbs (already in the past tense, for reporting) from which you should be able to select the exact shade of meaning required, from indefinite to very positive:

hinted (at)
implied, intimated
suggested
indicated, pointed to
showed, signified
demonstrated, specified
confirmed, substantiated
proved, established

Choosing the verb accurately should allow you to dispense with qualifying adverbs or phrases that lengthen and weaken the sentence.

importance of discovering observable phenomena of possible significance in petroleum exploration.

When asked what on earth that meant, the author interpreted confidentially:

This is a lousy set of data, but it may contain some information that will help find oil.

Bad news and the Persian Messenger Complex

"We cannot meet your deadline."
"We just lost the Megabux contract."
"I simply can't recommend the promotion of our vice-president's daughter."

The Persian messenger complex.

"That position has been filled."

"Sir, there's no way . . ."

With problems like these, writers often develop the Persian Messenger Complex—a dread that bearers of bad news may have their heads lopped off, as in the days of the ancient Persian emperors. And as those messengers must have done, the writer dawdles along. He leads up very gradually to the bad news. Or he buries it where it will be hard to find.

Since every reader and every situation is unique, the best way to present bad news should be determined through an extra-thoughtful analysis of both. In general, we can say on the basis of interviews with a great many typical readers that a long buildup to a letdown rarely works. On the other hand, a blunt-instrument approach doesn't work too well, either.

The question is whether the NEWS should be delayed at all, and if so, for how long? You might try setting up an imaginary telephone conversation with your reader. Assume he is on his way to a meeting, and you have only a couple of minutes to talk. How long could you stall?

The "messenger" who carried bad news in the following example would have profited from an analysis of the reader. Because her boss was anxious to hang on to all their in-house customers, she decided to delay as long as possible. We give below the original statement of the NEWS without informing you of the background, so that you can appreciate the effect of the suspense:

> In the limited time available for work on your project, the following progress has been made: [A list of seven items followed.]

Sounds pretty good, doesn't it? That list made about half a page of positive, cheerful information. It was followed, however, by a description of building renovations that had gutted the laboratory and delayed the project. At the bottom of the page, the in-house customer was informed:

> Considering your urgent needs, fairly small sample load, and the reasonable charges of contract labs, you might find it advantageous to send your samples to an outside service for analysis. I expect this solution would be cost-effective in the long run.

The subsequent revision of the draft, following a meeting with the boss and based on a genuine concern for the reader rather than the messenger, undoubtedly served both better:

```
INTRACOMPANY MEMORANDUM

To:   Mr. Fernandez, San Juan Office   Date: 12 July
                                               1989

From: Ms. Quemist, St. Louis Lab        Ref.: Sample
                                             Analysis
                                            Job #8907

Because an accelerated renovation program has
dismantled our lab, I must inform you that we
cannot complete your sample analyses this month. In
view of your urgent need, your fairly small sample
load, and the preliminary work we have already
completed, I suggest that we have the remaining
analyses done by a contract lab. I will be happy to
schedule and oversee this work for you.

The analyses we were able to complete before the
construction crews reached our lab include . . .
```

Most companies have recruitment form letters for upcoming graduates they have interviewed on campus. One of these is the rejection letter. It typically begins with thanks for the interview, then proceeds with a complimentary paragraph extolling the qualifications of the candidate and the high regard in which he is held. And then . . . regrets.

Graduates and new employees often feel quite cynical about such a delay of the NEWS. As one student put it, "The minute they start appreciating your qualifications, you know what's coming." It is not the opening thank-you they resent, nor the recognition of their superior qualifications; it is the wimpish delay of the eagerly awaited news—good or bad—that irritates them. This delay reminds us of the man who decided that his dog's tail should be docked. "I wanted him to get used to the idea gradually," he told a neighbor, "so I cut it off a little bit at a time." Spare your readers this kind of thoughtfulness.

Before you tackle a bad-news document, stop and consider: Is it *really* a problem for you as the author? If the answer is "yes," can you recast the bad news in a positive rather than a negative way?

Instead of "You will have to re-run five of the detailed calculations," try "Twenty of the detailed calculations are fine; we will need your help with the five others."

Reporting on an R&D failure

Realistic readers do not expect every research and development (R&D) project to be a howling success. In dealing with failures, partial failures, and partial successes, the object is precisely the same as with complete successes: to tell readers what they need to know.

This is easy to say but sometimes hard to do. Failure arouses the protective instincts of an on-the-job writer, because all that good **work** is truly not reflected in the bad **results.** Therefore it is easy to drift into a filibuster (delaying action) that postpones the inevitable news, or to scatter the news with bits and pieces of suggestive results throughout the document.

Readers are forced to reach tentative, changing conclusions as they go through such a report, only to learn at last that this exercise was futile: the project failed and has been abandoned.

The following tips may ease the task of reporting on an R&D failure:

- With few exceptions, the logical way to handle a failure is to report negative results as you would positive results: right where readers would ask for them if they were talking to you instead of reading.

- While there is no excuse for an annoying delay of the NEWS, there is no need to be jarringly blunt, either.

- The term *failure* is a harsh one. It can be avoided.

- A project that fails to meet its primary objective may still produce information of great value. It is a service to the reader to place this information where it will not be overlooked: early on. The fact that it mitigates the effect of bad news is a bonus.

- Never be unduly pessimistic about negative results. They are often used as legal evidence in obtaining a patent on a new concept. Unnecessarily negative reporting may well destroy any evidential value the report might have at a later date.

■ Avoid unwarranted adverse comments on the idea itself. It may yet prove to hold promise, even though the project to develop it has failed. A negative treatment might discourage future development.

Reporting the negative in a positive manner

If you have never had to face this problem, perhaps you simply haven't been around long enough. Many projects that appear (to outsiders) to be failures are only temporary setbacks. Without a positive reporting approach, you may be inviting all sorts of bad responses. Funding for the project might be withdrawn. A potential technological advancement might be lost. Large investments of time and money might be needlessly sacrificed. Credibility or goodwill might suffer undeserved damage.

The worst preparation for writing such a document is to misinterpret "honest but positive" as a euphemism for "devious" or "dishonest." We recall one lamentable case in which an ethical manager requested an equally ethical researcher to "write it honestly, of course; but keep it positive!" The researcher tried but could not strike the note his boss wanted. The upshot was a heated argument between them, followed by the resignation of this valued employee, delay of a promising investigation, and finally, the shelving of the project. Before it could be resumed by a replacement, a competitor had picked up the idea, developed it, and put a profitable new product on the market.

Only an analysis of readers can define a positive approach. One good writer came up with a surprising negative statement as a positive opening. He analyzed his most influential reader as an outspoken critic of big business, fond of voicing his opinions of "self-serving" business writing. In this situation the writer's progress was introduced in this way:

Despite the successes we have had in applications of product Y (see reports of May through August), we have recently found one application where its use is nothing less than disastrous. It cannot yet be made to. . . .

Obviously, not all failures can or should be handled in this manner. Another writer's analysis indicated that the fact of failure was actually less important to her readers than the problem that caused it. In this instance, treating the problem ahead of the result turned out to be a positive approach because she had a pretty good idea of what caused it and what to do about it.

Another writer had failed to reach his stated goal but had nevertheless made six unexpected, very positive findings. Thus he was able to draw seven conclusions, beginning with the negative one, followed by six positive ones that far outweighed the failure.

Anticipating antagonism or rejection

Here is another situation that invites a little delay of the NEWS. And that might be a good idea—certainly you would not approach a hostile reader as you would a friendly one. To illustrate: imagine a memorandum as it would be written to the two supervisors described below (see Fig. 7).

Situation: Supervisor has asked author to investigate available decontamination cabinets and to recommend the one best suited to the group's needs. Author has looked at cabinets manufactured by Companies A, B, C, and D. Only D is suitable.

Analysis of Supervisor #1: Had a bitter experience with a cabinet from Company D several years earlier. Wasted money, exposed some employees to contamination, and got into trouble with his supervisor. Recovered from that situation, but with enduring hostility toward Company D. Writer can anticipate an explosive "No!" if Company D's name is brought out of the blue. Considering this supervisor's initial mind-set, would you be wise to begin directly with your real NEWS—the recommendation that a Company D cabinet be purchased?

Analysis of Supervisor #2: Has an open mind. Will accept any justifiable and cost-effective recommendation. In this case it would be patently foolish to consider any delay of the NEWS at all. You would recommend the cabinet from Company D, tell why, and attach a requisition for signature.

INTEROFFICE MEMORANDUM

To: Sue Purvyser   Date: 12 November 1991

From: V. R. Coshus   Ref.: New Decontamination Cabinet

The attached table summarizes the results of our investigation of available decontamination cabinets. In brief:

- Company A has discontinued the only models that would meet our requirements.
- B Corporation's cabinet would be marginally suitable, but the cost is prohibitive.
- C Company model's particle trap is inadequate for our work, and modifications would double its cost.
- The only cabinet that meets our requirements is a popular model introduced in 19____ by Company D. Unlike previous Company D cabinets, Model 1499T has safety features that . . .

We have interviewed five users of Company D's cabinet, all of whom recommend it. We therefore recommend the purchase of Company D's Model 1499T at a cost of $3,295.00. A purchase order is attached for your signature.

INTEROFFICE MEMORANDUM

To: D. Bosman   Date: 12 November 1991

From: Z. Undorling   Ref.: New Decontamination Cabinet

Attached for your signature is a purchase order in the amount of $3,295.00 for a Model 1499 T decontamination cabinet from Company D. The table below, summarizing the results of the investigation you requested, shows that the recommended model is clearly the best choice for the following reasons:

- It has a high-quality particle trap that will meet our needs (the trap on the C Company model is entirely inadequate).
- It is less than half the cost of B Corporation's comparable cabinet.
- It is only slightly smaller, and significantly cheaper, than Company D's Model 1699U, which . . .

Figure 7   A memo in which the NEWS had to be postponed *(top)* and a straightforward memo that will minimize the supervisor's effort *(bottom)*.

Need for tact or diplomacy

Have you ever been obliged to do any of the following?

Explain to management a bad decision, costly error, or some other action for which your supervisor was responsible?

To avoid finger-pointing, describe reasons rather than decisions made. If that is not possible, prefer passive verbs: "The decision was based on . . . ."; "It was decided that . . . ." (True, we normally ought to avoid these circumlocutions; but here we have a special problem.)

Dispute the findings or criticize the work of a colleague in a letter to his management?

Make positive suggestions rather than negative criticism.
This: "Before implementing the Jones program, we should determine . . . ."
Not this: "Jones' program fails to take into consideration . . . ."

Advance your view in opposition to another that is so stupid it makes you angry?

After dictating the rough draft, back away; give yourself a chance to cool off. Return to your message and edit vigorously. It is impressive how much more objective you can be once your anger has been vented in writing.

Tell the boss's boss that he is wrong?

Begin by acknowledging the logical aspects of his reasoning before introducing—carefully—the illogical and erroneous aspects.

Those are only a few of the situations that call for tact or diplomacy. Each is unique, requiring its own approach. But if you can't decide before writing what that approach will be, including your introduction and carefully chosen wording, your document will probably be disorganized and it may be offensive as well.

Technical conflicts (and bias)

A writer complains:

We don't agree on technical aspects. If I write it my way, my supervisor won't approve it. If I write it his way, I don't even want my name on it!

Once that is out in the open and out of his system, the typical writer simmers down; he knows he cannot always have the last word. His chief problem, then, is how to write without exposing his bias. He might actually enjoy sneaking in a bit of personal bias for his own satisfaction—and for the benefit of his readers, as he would perceive it—but that would only lead to a rewrite.

Unintentional bias reveals itself in such subtle ways that it may be hard to put one's finger on the exact wording, sentence structure, omission or commission responsible for its presence. But that presence will be sensed whether or not it can be explained.

Media bias (intentional or unintentional) illustrates the way in which the mere choice of verb or modifier can express feelings that media professionals supposedly try to keep out of their work. For example, a national TV newscaster showed the following printed statement on the television screen (we omit names):

Dr. Z stated that this is a Type W theory.

But as the newscaster quoted that statement aloud, he edited it as he went, like this:

Dr. Z ~~stated~~ admitted that this is a Type W theory. *only*

The change in verbs from "stated" to "admitted" made Dr. Z somehow guilty of something, while the insertion of the modifier "only" belittled Dr. Z's theory.

While technical and business writers may have equally strong sentiments, they are not likely to make such an obvious blunder. Their bias typically creeps in through sentence structure—chiefly incorrect subordination, which leads to upside-down organization within the sentence.

There was, for example, an employee benefits specialist who had investigated three alternative programs and had arrived at the only sensible choice (in his opinion), which we'll call alternative B. He was obliged, however, to recommend alternative C, for that was the choice of his two supervisors after they reviewed his data. After a losing battle, his rewrite introduced the alternative he preferred in this way:

Although Program B would entail a greater expenditure, it has certain decidedly superior features.

The author's interest (and bias) are reflected in the "skeleton" (underscored words) of the main clause*. Since that skeleton always sets the theme for a text unit, the writer instinctively completed the paragraph by extolling the virtues of alternative B. The subordinate clause referring to the supervisors' main interest (the cost) got short shrift; it was the last thing discussed in the unit.

One of the supervisors rewrote the whole section and established her own theme by reversing the subordination in the lead sentence, as follows:

> Although alternative B has certain superior features, its <u>cost</u> <u>would</u> <u>be</u> <u>prohibitive</u>.

The underscored skeleton of this sentence sets an entirely different theme, and the ensuing discussion was centered on prohibitive cost rather than superior features. Although the two lead sentences contained the same information—cost and features—the effect was very different.

A matter of ethics

More than a few authors in our workshops, especially relatively inexperienced professionals, have brought up the subject of ethics when they have been assigned a subject and a viewpoint (position) on it that they do not personally hold. "Signing my name to this paper," they say, "would be dishonest."

Their problem deserves sympathetic handling. Employees who care about integrity are a valuable asset to any organization. Steps can be taken to relieve the conflict between their loyalties and ambitions on the one hand and their sense of what is ethical on the other.

Nevertheless, decisions that affect the operations of an organization are necessarily made final at a higher level—a vantage point from which the overall view is broader. The chain of command cannot be dismantled for the benefit of individuals who may disagree with policy decisions. The question would then become, Who runs this outfit?

If you could visualize yourself in the position of a professional editor or ghostwriter, you would have no problem articulating what somebody else believes. Then you could ask a supervisor to

---

*Sentence skeletons are explained in Part III, p. 170.

sign the document. Chances are the supervisor would not mind taking credit if you had done a good job. If that fails, here are some options:

- Propose transferring the job to a co-worker with a better background and feeling for the job.
- Explain and refuse, regretfully. A people-oriented supervisor will respect your motives. But in case you don't have that kind of supervisor, perhaps you should be prepared to . . .
- Look for another job. The problem with this is that you are likely to run into the same problem again.

(A last, possibly flippant suggestion: Have you thought of going into business for yourself?)

Having to follow unsuitable guidelines

As we mentioned in Part I, prescribed outlines may not be at all suitable for your subject. While many professional journals now permit contributors to exercise some judgment regarding the format of professional papers, others still furnish editorial guidelines that require a standard pattern such as **"Abstract," "Introduction," "Background," "Methods and Materials," "Discussion of Results,"** and **"Conclusions"** or **"Summary"**—in other words, some variant of the antiquated Suspense Format.

Arbitrary guidelines sometimes fit the author's material and purposes, and also the reader's interests. But when they do not fit, they simply lengthen the paper and raise production costs unnecessarily. According to *Economics of Scientific Journals* (1982), if one article of 24 pages could be cut by a third, it would save $160 to $800 in publishing costs.

To illustrate the problem, suppose you are an ecological scientist writing an article for your professional journal, a special edition devoted to the effect of various industries on the preservation of North American forests. This hypothetical article deals with the newsprint industry.

The outline specified by the journal requires a section entitled "Analytical Methods." This does not fit your material. Version 1 of your article shows what strict adherence to the outline might produce:

Version 1

Analytical Methods

Total U.S. newsprint demand in 1988 was estimated assuming a 2.7-percent increase in usage, offset in the West by a 100,000-ton drop in inventories, and augmented in the East by a 250,000-ton increase. Estimates of consumption by U.S. daily newspapers were based on a 0.8:1 relationship with GNP growth. The consumption figures do not include the market segment represented by applications such as supplements and flyers, assumed to be growing 50 percent faster than GNP. Other factors excluded . . .

The data on which this analysis was based were supplied by the Continental Council of Newsprint Manufacturers, covering a 10-year period . . . .

Why not indulge in a bit of literary license under that "Analytical Methods" heading? Tie **methods** to **results.**

Version 2

Analytical Methods

Analysis of statistics for the period 1979–88, which were supplied by the Continental Council of Newsprint Manufacturers, indicates that the demand for newsprint is heading for new highs in the next decade. Projections into 1997 (Fig. 1) predict a serious effect on the preservation of American forests unless new fiber utilization is developed.

The increase in total U.S. demand for newsprint over the last decade is estimated at 34 percent. This estimate assumes a general 2.7-percent increase in usage in 1988, offset in the West by a 100,000-ton drop in inventories, and augmented in the East by a 250,000-ton increase.

Consumption by daily newspapers alone, reflecting a 0.8:1 GNP growth, has climbed at an average 2.9 percent per year. This analysis of consumption does not take into account the segment of the market represented by advertising supplements and flyers, which has been growing 50 percent faster than GNP. Other factors not included are . . .

Nobody can claim that "methods" are not there. They are

there. But they are not crammed into paragraphs in which their relationship to conclusions is lost. They will not have to be introduced again during a discussion of results.

In fact, if you were free to compose suitable headings of your own, that tired old "Analytical Methods" could usefully be replaced by something like "Newsprint Demand Climbing." You could also eliminate several mandatory (but unneeded) parts of the prescribed outline, automatically reducing length.

Here are a few more ways to get around the unsuitable outline:

- If you cannot get to the point with your NEWS immediately in the body of the paper, you can certainly do it in the abstract or opening summary.
- To reduce length, you can at least get to the main point of each unit within the text, no matter what its heading. Here is an example from a geological field report in which the standard heading "Topography and Access" was required:

Topography and Access

The area is accessible only by helicopter except in the spring, when the Flujo River is navigable. [That is the real NEWS within this section. An explanation of WHY will then lead into a description of the topography, which in this instance is largely of academic interest.]

- If you have to dredge up material for a compulsory unit heading, the least you can do is keep it brief. Guidelines never require you to make it formidably fat.

Getting the right answer

How often have you asked someone for information, only to receive an answer that was completely off target? Here are ways to ensure that you get the response you need:

(1) Make your request as concrete and as straightforward as you can.

(2) Use the imperative (command) form of the verb, as we have done here, preceded by "Please" or "Would you please."

THIS	NOT THIS

| Not later than May 23, please send me (Room 505) a tabulation of all visits that MetroTwix economists have paid to our office this year, with a one-sentence summary of each decision reached during each visit. | It would be greatly appreciated if a detailed report could be obtained in the near future on the recent meetings with Metro-Twix. |

(3) If you need a highly structured answer involving a lot of information, take the time to lay out a form that will require your respondent to feed back information in the sequence that is most useful to you. For example:
   (a) Frame your questions so that they can be answered "Yes" or "No." Leave a space for the answer.
   (b) Ask questions that can be answered by a number, a date, or other brief response, and leave the blank to be filled in.

Besides being easier to work with, answers obtained in this way help keep files from becoming overstuffed with useless material. Your investment of time will pay dividends later.

Multiple NEWS items and complex audiences

Multiple independent items of NEWS and audiences with highly varied interests are actually special problems in the organization of the message. You will find them discussed in Step 2 (pp. 64–67).

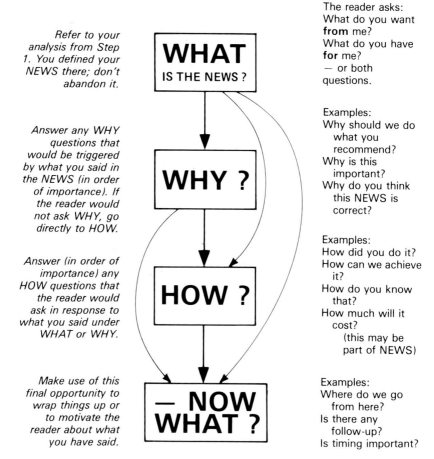

Refer to your analysis from Step 1. You defined your NEWS there; don't abandon it.

WHAT IS THE NEWS ?

The reader asks: What do you want **from** me? What do you have **for** me? — or both questions.

Answer any WHY questions that would be triggered by what you said in the NEWS (in order of importance). If the reader would not ask WHY, go directly to HOW.

WHY ?

Examples: Why should we do what you recommend? Why is this important? Why do you think this NEWS is correct?

Answer (in order of importance) any HOW questions that the reader would ask in response to what you said under WHAT or WHY.

HOW ?

Examples: How did you do it? How can we achieve it? How do you know that? How much will it cost? (this may be part of NEWS)

Make use of this final opportunity to wrap things up or to motivate the reader about what you have said.

— NOW WHAT ?

Examples: Where do we go from here? Is there any follow-up? Is timing important?

Figure 8 Step 2—How to develop the Essential Message.

Step 2
Develop Your Essential Message

> I feel strongly that ideas which cannot be made clear to an audience are not clear to the writer . . . the question is perfectly legitimate: "What is the *point?*" Every paper and every book should have a point.
>
> —Lester S. King, MD*

**USE THE PATTERN
OF CONVERSATION**

The most consistently successful method yet found for communication is face-to-face conversation. If you want something from a colleague, you ask for it; if you have information he can use, you offer it. That is your NEWS, and it controls the ensuing dialog. It triggers a series of questions that you answer as they arise. Dialog is that simple, that direct. It is fast, efficient, and economical.

Fortunately, the Q&A pattern of conversation is quite predictable, as represented in Fig. 8, so it makes a simple, effective, and realistic guide for organizing nonfiction writing of any length or degree of complexity: letters, memos, reports, speeches, professional papers—even minutes of meetings.

The use of this guide is simplified by the fact that the hardest part of writing has already been taken care of in Step 1. You have already defined the NEWS, which calls forth the readers' questions you will be answering. You can anticipate and frame those questions accurately because Step 1 has also identified your readers' interests.

Organizing by this method is a natural, familiar act. You are drawing on a skill that you began acquiring in childhood and have been practicing throughout your adult life. In conversations with colleagues, friends, and family, you demonstrate daily that you have the ability to communicate with real craftsmanship. You can achieve the same degree of excellence by applying your conversational skills to the organization of written documents.

---

*Lester S. King, MD, and Charles G. Roland, MD, *Scientific Writing,* AMA Publications, 1968, p. 67.

HOW TO DEVELOP THE
ESSENTIAL MESSAGE

Taken together, the answers to the questions in Fig. 8 form the framework, the ESSENTIAL MESSAGE, for any kind of informative writing. Because the questions are based on an analysis of interests and objectives (Step 1), they tailor each document to fit the needs of specific real people—not nebulous, hypothetical, or standardized "readers" or "audiences."

Further, focusing on that individuality forces us to think in terms of practical use, economics, timing, suitability of language and tone, need for qualification and definition, and requirements for detail. In other words, with Steps 1 and 2 you custom-engineer the product.

Now let's take a closer look at the guide questions of Fig. 8.

What is the NEWS?

You can count on the NEWS to be Interest #1 nearly 100 percent of the time. About 98 percent of the time you can get to that point immmediately, without any buildup (Statement of Problem; Background; Objectives; Methods; Data; Discussion; etc.).

The Bottom Line

The business term "the bottom line"—literally the final line of figures in a financial statement—has spread into popular use for referring to the overall outcome, the principal result, the net effect. When the boss says, "Just give me the bottom line," he means: "Never mind the details. What's the main point?" So, when we get rid of the Suspense Format in our writing, the **bottom** line becomes the **top** line.

There are times, of course (the other 2 percent), when one cannot or should not be so direct. A brief delay would be logical if the subject required definition before it could be discussed; the reader needs to be able to follow the discussion knowledgeably. As mentioned earlier, certain special problems can also justify a delay.

Remember, nevertheless, that delay can be counterproductive.

Here is a way to decide: withhold your main point no longer than your reader would tolerate the suspense if he were present and could interrupt with questions.

Don't Just Make Promises

Don't confuse a promise of news to come with the NEWS itself. Some writers offer a teaser and think they have met their obligation to expectant readers.

In Fig. 9, the first news headline is a teaser, a promise that you will eventually come upon some hard facts. It is much like the opening sections of many technical reports, and it is typical of a high percentage of scientific abstracts that promise:

> In this report, so-and-so is analyzed; such-and-such is reviewed; for the first time, so-and-so is explained; finally, this detailed investigation has made it possible to draw conclusions about such-and-such.

Such abstracts, called descriptive abstracts (Fig. 4A, p. 22), are reminiscent of the old girly shows in a travelling carnival, which revealed to the rubes outside just enough to get them to buy an admission ticket.

Figure 9 Which headline contains real NEWS?

In one of our classes a crusty old plant supervisor, resisting the upfront treatment of his NEWS, complained, "Dang it, you're askin' me to tell my punchline first! That ruins my story!" One of his colleagues set him straight: "When I want to be entertained, I won't be reading your 'story,' buddy; I'll read a murder mystery or go to the movies. Don't pull your punches with me—hit me between the eyes with your NEWS."

Why . . .?

Examine your statement of the NEWS. If you were conversing with your readers, would they ask WHY in response to that statement?

WHY should I care—what's in it for me?

WHY did you do that?

WHY should we do that?

WHY do you want it? propose it? recommend it?

Answer each such question with a **complete** sentence.

The WHY section is the obvious place to justify or sell, to explain advantages, needs, reasons, economics, safety benefits. To highlight selling points here, separate them with white space—a listing in which each point is introduced by a bullet (●) or dash. If you need to refer to the points later, or if you wish to emphasize their ranking, number them.

Don't get hung up on guide questions

If you cannot foresee any WHY questions, don't try to answer any. Always keep in mind that the four questions in Fig. 8 are only a guide; they are not a rigid sequence to be followed in every document. Only people incapable of original thought follow rules blindly.

Few of your readers would ask WHY if you told them there was a new, entirely legal way to cut their income taxes in half. Their response to that NEWS would be "HOW can I cut my taxes in half?"

On the other hand, you could anticipate a chorus of WHY's if you informed readers of this:

Beginning next January, the IRS will require all taxpayers

to make an early installment payment of $2,000, whether or not they will owe taxes at the end of the year.

It is often easy to rephrase a WHY question and turn it into a HOW (and vice versa). For example, in response to a surprising NEWS statement, one writer might phrase the reader's question as "WHY do you say that?" while another writer might put it this way: "HOW do you know that?" In either case, the same answer would be forthcoming—in the proper sequence.

Sometimes you may anticipate that readers would ask a HOW question before a WHY question ("HOW much will it cost us?"). If that is the logical order, by all means reverse the sequence in Fig. 8.

How . . .?

Refer to your previous statements in response to WHAT and WHY questions. One or more HOW questions might follow:

> HOW do you know that?
>
> HOW did you do it?
>
> HOW should we do it?
>
> HOW did it happen?
>
> HOW much . . .?
>
> HOW many . . .?
>
> HOW about alternatives?

Here is the appropriate place for the technical detail with which many professionals feel most comfortable: the work, the data, the results. Sometimes when writers are introduced to the conversational approach, they become uneasy, fearing that the Murray System will deprive their readers of the nitty-gritty details.

It won't. If the questions readers would ask in conversation are answered, there will be no sacrifice of desirable content. Discussions, data, test results, etc., will fall naturally into place **in support of main points.** The method does discipline the writer, however, to omit unnecessary information and *superfluous* detail, thereby limiting document length and at the same time improving readability.

There is an additional opportunity to reduce length and improve readability by **getting directly to the point at every major level** within the HOW section. For instance, if a report is divided into several chapters, each of which has a number of sections with side headings, all those first- and second-order topics can be introduced with their own NEWS before the supporting detail.

Let's take the example of a writer whose report contained a section concerning 12 tests that were run to determine what caused subsurface formation damage in a wildcat drilling operation. *Twelve* tests! That meant he got his answer the hard way. His readers got it that way, too, for he concentrated immediately on the labor instead of the product:

> Twelve <u>tests</u> <u>were run</u> to determine what caused the formation damage.

Since the skeleton of the sentence (underscored) established the theme for discussion, the author automatically followed that lead with 12 test descriptions and 12 individual test results.

But the tests were history. The author knew the answer. His readers needed the answer. He should have given it to them straight away:

> Twelve <u>tests</u> <u>demonstrated</u> that the <u>materials</u> used in completing the <u>well</u> <u>caused</u> the formation <u>damage</u>.

We tried that introduction in a rewrite, and were rewarded with a 33 percent reduction in length without loss of information.

Back to Fig. 8.

Now What?

Your option: If there is nothing more to say, this is a dandy place to quit. Don't feel that you are obligated to taper off with a "parachute" section that will let the reader down gently. However—

- Readers may have some final questions: What will you do next? What should they do next? Where does this leave us? What's the timing on the next phase?

- You may want to tie up a neat package of important points if the document is long or the concepts difficult. (For short documents, the opening summary suffices.)

- Since the last thing we leave our readers with has almost as great an impact as the first thing we hit them with, the NOW WHAT section is a good place to motivate readers to take action, to sign an attachment, to meet a deadline. Or we can just use it to leave readers feeling the way we want them to feel about what we've said or done.

- In letters or memos to persons we know well, or in documents that ask favors or that reply to favors, we often like to add a personal note or an expression of appreciation. A gracious ending is always appropriate in these circumstances. Just be sure that your final sentence or paragraph is not a trite sign-off, like the "please feel free to call on me" with which an insurance analyst ended a memo to her own boss. (Another thoughtless letter-writer even added that cliché invitation to a letter he had carefully crafted to put an end to unwelcome correspondence!) Parachuting can become a habit.

HOW TO USE THE ESSENTIAL MESSAGE

When the pertinent guide questions have been answered, you will have a brief but extremely useful Essential Message. Following are some of the ways you can apply it in different kinds of writing jobs.

Short Documents

The Essential Message may constitute the entire short memo, letter, abstract, or transmittal letter, in which case you can skip ahead to Step 4 or even Step 8. In some cases you may need to expand each statement into a paragraph; just be sure that your Essential Message statement introduces the paragraph. Readers should be able to scan the document, pick up lead sentences, and get the essence of the entire document—the story line, so to speak.

Long Documents

If you are starting a long document such as a formal report or professional paper, the Essential Message serves as a reader-oriented guide for an outline (see Step 3). It also gives you the essence of an excellent summary or abstract.

Note that we are reversing the old theory of organization, which held that one could not summarize until one knew what one had said. In other words, how do I know what I'm going to say until I see what I've said? The Essential Message provides a much more thoughtful way of planning a report; the fact that it also constitutes a concise summary is fortuitous—a bonus.

Impromptu Talks

Participants in some of our classes have reported that the discipline of putting the NEWS at the beginning (and then supporting it logically) has helped them a lot in off-the-cuff speaking. One of the most difficult things to do well is to stand up with no prior preparation—perhaps even without any warning—and speak in a logically organized fashion, especially if the subject is technical.

But with practice in composing Essential Messages, a speaker can transfer that skill to the quick mental organization of an ad-lib speech. If the speaker is nervous, the guide questions serve as easily remembered prompts. Once the main point has been stated, it is easy to support it by answering WHY and HOW questions. Before sitting down, the speaker restates the main point to drive it home. He thus comes across as a well-organized thinker, even though he may be a bit shaky as an orator.

Engineered Speeches

To find out how the Murray System applies to the planning of a speech, see Part V, p. 251.

Recording Minutes of a Meeting

Anyone who has taken the minutes of a meeting knows how hard it can be to capture the highlights of discussions. It is especially difficult if the chairman is poorly organized or if he allows simultaneous cross-conversations among participants. The form shown in Fig. 10 will help keep you on track.

The five column headings will prompt you as to what to listen for and what to fill in. Meetings with a lot of cross-talk typically

MINUTES OF THE Northern Division Safety Committee		LOCATION: Ms. Abel's Office	DATE: 2-16-87, 2 PM	
	(WHAT is the NEWS?)	(WHY? and HOW?)	(NOW WHAT?) IMPLEMENTATION	
AGENDA ITEM	DECISIONS, CONCLUSIONS	REPORTS AND DISCUSSION	WHO ?	WHEN ?
1. Fire at the Maple Trail Plant	Heat detectors will be installed in all Northern Division plants to supplement the present smoke detectors.	Mr. Baker reported that the 2/13 telex concerning the fire was essentially correct. Ms. Abel asked if any personnel were still in hospital; Mr. Conklin said only one man was still hospitalized with burns to his arm and hand. Ms. Dean reported the smoke alarm never went off because.....	Ms. Dean to prepare AFE and contact Purchasing.	2/19
2. Use of seat belts in Company cars	Will talk again to the Randle Plant union about supporting the seat-belt campaign. Will circulate a special bulletin to all employees emphasizing the seriousness of the new rule on seat belts.	Mr. Ethridge reported that the shift boss at the Randle Plant who was thrown through the windshield of his company car during a head-on collision is still in critical condition. Ms. Abel commented that the union had resisted Company pressure to get employees to buckle up and now the Company's public image as a safety-conscious organization had been damaged. Ms. Farley noted that . . .	Ms. Abel	--
3. Semiannual report on pla... ...ams	No decision was reached	Mr. Gordon reported that his group's part of th...	Mr. Ethridge	2/27

Figure 10 Format for minutes of meetings.

get bogged down in the WHY and HOW section, where it is possible to waste time recording a great many irrelevant comments. Even the chairman can get distracted and fail to realize that **no clear decision has been reached** on the point under discussion. Since decisions are the main product of any business meeting worth holding, a blank in the second column will alert you to the fact that none has been agreed on. (Occasionally you will end up recording a formal decision to postpone any decision; but even that is worthwhile.)

Once you have a decision, the conversational prompts WHO and WHEN in the last two columns will remind you to fill in this important information. Again, you may be able to help a disorganized chairman by asking for the names and dates required if they have not been clearly specified.

Having used this form to record the gist of a meeting, you have two choices: you can either transcribe your form-notes into a narrative account of the meeting or you can reproduce the format of Fig. 10 for your report. Some executives prefer this tabular format because the highlights can be read and digested quickly; specific agenda items (the left-hand column) can be located easily; and the two right-hand columns provide an action or "tickler" for future reference or calendar notations. (If legal considerations require a verbatim transcript as minutes, take a tape recorder or a good shorthand stenographer to the meeting.)

PROBLEMS THAT MAY ARISE IN STEP 2

As mentioned earlier, developing the Essential Message may be complicated if there are several NEWS items, or if there are two or more readers with different interests or varying degrees of interest. Solutions to these problems are discussed below.

How to Handle Multiple NEWS Items

Suppose you have completed a study and have arrived at three conclusions, each of which qualifies as an important NEWS item. Suppose further that your analysis of readers indicates that each statement of NEWS would elicit an immediate WHY question.

Would you follow the guide questions as shown on the left side of Fig. 8? That is to say, would you write sentences or paragraphs for *all* your news items (conclusions) and then follow with a second sequence of corresponding WHY answers? And if each WHY answer triggered an immediate HOW question, would you take those up in a corresponding third unit of text?

If you did, you would have split topics that would force the reader to jump back and forth, trying to find and match the related ideas, as represented in Fig. 11. Instead of neatly separated "oranges, apples, and bananas," you would have the written equivalent of a fruit salad.

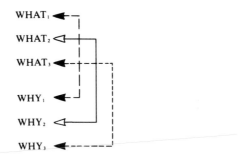

Figure 11   How not to handle multiple NEWS items.

To avoid a fruit salad of split topics, you would choose to combine WHAT, WHY, and perhaps HOW for each independent item of NEWS, keeping all except the NEWS subordinate. For example, here is a combination of WHAT and WHY:

> To provide care for infants in whom conventional feeding would be poorly tolerated (WHY), we have developed an improved nasojejunal tube-and-insertion technique (WHAT).

If desirable, a brief HOW response might also be included in the combination:

> Tests on 236 sick infants (HOW) have demonstrated that an improved nasojejunal tube-and-insertion technique provides an effective means of feeding patients (WHAT) in whom conventional feeding would be poorly tolerated (WHY).

If the same reply to WHY would apply logically to all items of NEWS, it could effectively be used to introduce them:

To determine whether age affects the sick infant's ability to tolerate the tube-and-insertion feeding technique (WHY), I recommend the following:

- WHAT (first recommendation)
- WHAT (second recommendation)
- WHAT (third recommendation)

How to Handle a Complex Audience

On-the-job writers are often uncertain where (and how much) information should be introduced in documents written for two or more people with different interests or degrees of interest.

Letters and memos present no great problem, because the message is directed to the addressee. Those who approve the text or receive copies are served by HOW paragraphs, attachments, and cover memos.

In fact, articles and reports should present no big problem either, for the normal divisions of summary, body, and appendices are ideally suited to serve the three general classes of readers: managers, technical peers, and specialists, respectively. We can satisfy all while burdening none if we recognize that audience decreases as detail increases (Fig. 12).

Here are some general rules, then, for designing reports for complex audiences.

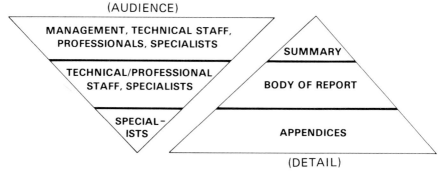

Figure 12   Breadth of audience decreases as detail increases.

(1) Design the summary primarily for management

Refer to your analysis of management interests in Step 1. The overview (the Big Picture or Bottom Line) is important to managers. They need the answers to WHAT, WHY, HOW, and NOW WHAT. Although they may read entire lengthy reports, don't depend on it. Often they will read the summary and then pass the report on to an assistant for comments.

While the summary is tailored to management needs, it serves the entire readership. It provides all readers with a perspective from which to view details with understanding.

(2) Design the body of the report for your technical peers

By the time the body of the report is reached, you may have lost management readers. Now you are talking to people who need sufficient information to determine the accuracy, reasonableness, and utility of the work or ideas.

How much detail? What kind? Again, refer to your Analysis of People. Don't make the mistake of assuming that your peers are all eager for as much detail as a proud author would wish. A major corporation with worldwide affiliates and a vast potential readership found less than one-half of one percent interested in the kind of detail we usually (and with good reason) relegate to appendices.

(3) Design the appendices for specialists

These are the readers whose interests can be assumed to be more nearly like your own. The specialist might want sufficient detail to repeat your experiment; to examine raw data; to try a new technique; to check the derivation of your equations; to run your computer program; to study volumes of historical minutiae.

An appendix is the place for material that would repel the majority of readers in a complex audience. It is also a good place to preserve information you think "somebody might need, some day" when you aren't sure. If you must make it available, be certain that it doesn't clutter up the body of your report or article. Better yet, write a supplementary report (see p. 87) and offer it to readers.

**DEVELOPING THE ESSENTIAL
MESSAGE: TWO EXAMPLES**

Example 1—A Short Document

The Situation

Florence B, a professional in the Systems Group of a large nonprofit institution, was writing a simple instructional memo requiring very little expansion beyond the Essential Message. Her subject was the availability of six powerful lap-size microcomputers for which she was responsible. Her boss asked her to circulate instructions to executives who might take the machines on trips. Ms. B was able to do Step 1 in her head, as follows:

Florence B's Analysis of People Involved

(1) WHAT do you OFFER?
New Model P computers for our executives
(2) WHAT are readers' interests?
Addressees: 17 executives
(a) Why should I use one of these computers?
(b) How do I get hold of one?
(c) How do I use it?
From (signer): Manager, Systems Group
Same interests as those of Ms. B.
(3) Do you anticipate any special problems in writing?
No.

Florence B's Essential Message

WHAT is the NEWS?
Here are instructions for obtaining and using the new Model P briefcase portable computers that are now available to you for out-of-town trips.
WHY (should I use one)?
On the road you can use the Model P for financial analyses, word processing, and communication with our mainframe computer at headquarters (see Attachment 1 for details).
HOW_1 (do I get one)?
(1) Since we have only six of these units, call ahead (Ext. 2990) to reserve one.

(2) The day before your trip, pick up your Model P at Room 314 South in order to check it out.

HOW$_2$ (do I use it)?

(3) The lid of the carrying case contains a manual with complete instructions for the use of the Model P.

(4) Before leaving on your trip:

    (a) Run the diagnostic test for battery condition (see manual, p. 17).

    (b) Plug the modem into your office phone line and dial 369-1212 for a check of the built-in communications program.

    (c) Make sure you know how to load and use the built-in spreadsheet and word-processing programs, and how to load any demonstration programs you plan to take along (Attachment 2).

(5) If you have trouble with the instructions, call Florence B (ext. 2990).

NOW WHAT?

At the end of your trip, please fill out the evaluation form you will find in the lid of the computer case and return the form to Room 314 South, along with the computer.

Note: Using this Essential Message, Ms. B dictated the full text of her memo, supplying explanations as needed following each statement. As Fig. 13 shows, very little expansion was required.

Example 2—A Long Document

The Situation

Natural gas company Vice-President M requested Law Department Manager A to determine whether NGL (natural gas liquids) temporarily stored in company tanks was taxable by the port city in which the tanks were located. Ms. A delegated the job to Mr. G, a tax lawyer.

Mr. G researched applicable laws, reviewed files, and made trips to the facility in question and to similar ones in other places. He determined how the NGL was acquired, where it came from, the purpose of storage, how long the NGL remained in storage, and how taxes had been handled previously at the various sites.

Back in the office, Mr. G studied old and new property tax codes, definitions, and previous court decisions. He also reviewed

┌─── INTEROFFICE COMMUNICATION ────────────┐

To: Attached Distribution List   Date: 3 March 1987

From: Manager,   Ref.: New Model P Computers
      Systems Group

Here are instructions for obtaining and using the new Model
P briefcase portable computers that are now available to you
for out-of-town trips. On the road you can use the Model P for
financial analysis, word processing, and communication
with our mainframe computer at headquarters (see Attachment 1 for details). As these units are part of the Company's
experimental PIP (Productivity Improvement Plan), there
will be no in-house rental charge for their use until further
notice.

Instructions

(1) Since we have only six of these units, please call ahead
    (Ext. 2990) to reserve one. They will be made available on
    a first-come, first-served basis.
(2) The day before your trip, pick up your Model P at Room
    314 South and check it out.
(3) The lid of the carrying case contains a manual with com-
    plete instructions for the use of the Model P. There is also
    a handy condensed instruction card for those already
    familiar with the machine's operation.
(4) Before leaving on your trip:
    (a) Run the diagnostic test for battery condition (see
        manual, p. 17).
    (b) Plug the modem into your office phone line and dial
        369-1212 for a check of the built-in communications
        program. Our testing showed occasional problems
        with this program.
    (c) Make sure you know how to load and use the built-in
        spreadsheet and word-processing programs, and how
        to loan any demonstration programs you plan to take
        along (Attachment 2):
(5) If you have trouble with the instructions, call Florence B.
    (Ext. 2990).
(6) At the end of your trip, please fill out the evaluation form
    you will find in the lid of the computer case and return
    the form to Room 314 South, along with the computer.
    Your feedback will help us determine which programs
    are most useful and whether we can justify obtaining
    more of these units.

└───┘

Figure 13   Example of a short memo derived from the Essential
Message.

the trip reports and numerous memos to file that he had written during the course of his investigation.

In view of the considerable investment of his time and effort, and with all the details of reports and memos as source material, he could easily have been led into a traditional suspenseful report. He wasn't. He analyzed the interests of the people involved.

Lawyer G's Analysis

(1) WHAT do you have to offer?
    Answer to V.P.'s question: No. (NGL not taxable by City X)

(2) WHAT are readers' INTERESTS?

People Involved	Interests
Addressee: V.P., Mr. M	Is the NGL stored in tanks at _____ taxable in City X?
From (signer): Manager, Ms. A	Is the NGL taxable by City X? Validate the legal opinion.
Through (approval): Section Head, Mr. T	Same as above, plus background, history, court decisions, etc.

(3) Do you foresee any special problems in writing?
    Yes: how to give section head all the details he insists on without burdening V.P. and Manager.
    Strategy: Answer V.P.'s question first; provide details for V.P. and Manager next; put details for Section Head in attachment.

Lawyer G's Essential Message

WHAT is the NEWS?
    After reviewing the applicable facts and law, we find that the NGL in Company storage tanks in City X is not taxable by that city.

WHY (is it not taxable)?

Our NGL enters working storage from a common-carrier pipeline and remains in storage for no more than seven days. Under state case law, NGL in a common-carrier pipeline is in transit and is thus located there temporarily.

Under the statutes of this state, property is taxed where located only if located there more than temporarily, which is defined as no longer than 180 days before it is permanently removed.

HOW (do you know that)?

(Lawyer G cited laws, statutes, and regulations; provided a map pinpointing the storage area; etc.)

NOW WHAT?

(Omitted; not applicable.)

Note: Lawyer G's memo was plain, direct, and complete. With the attachments, it contained all the information all readers wanted. This memo would have met the requirements of the "Plain English Movement" as described in the November 1983 *Michigan State Bar Journal*. (That issue of the *Journal* includes a bibliography of 146 articles and books on "Plain English for Lawyers." We recommend it to all lawyers.)

THE NEXT STEP

If you are writing a short document, the Essential Message will usually be all you need to finish the job (except for Step 4 if you have illustrations). The Essential Message *may* constitute the entire document; in that case, go to Step 8—Edit. Or you may need to expand the text slightly. If so, go to Step 7—Dictate.

If you are writing something long enough to require more detailed organization, go to Step 3.

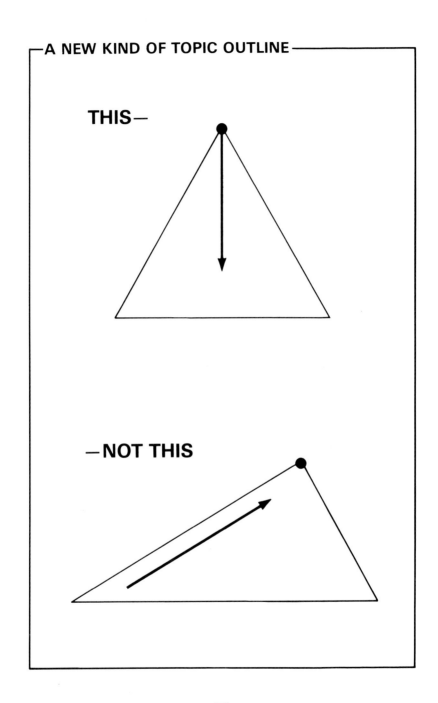

Step 3
Construct A New Kind Of Topic Outline

Writers who are entrenched in the suspense method of outlining they learned in college find it difficult to imagine any other way of organizing ideas. Surely turning the body of a report upside down would bury their good work, diminish their prestige?

On the contrary; readers consider a report outline based on their own interests instead of the author's pride to be quite acceptable. Their opinion of the author, as both worker and writer, is based primarily on his results and only secondarily on his techniques. We can admire a well-planned and -executed project, but we can *use* good results. And if results were not good, the author's prestige is not enhanced a bit by hiding or delaying that NEWS in a lengthy discussion of his labors, however brilliant.

The Essential Message (Step 2) arranges main ideas as readers would call for them in conversation. Using that message as a guide for a conversational outline, then, ensures that the body of the report will follow the same reader-oriented sequence. Main points will stand out. The technical sections of the report can be scanned quickly for their essence.

Moreover, preparing the conversational topic outline is a valuable discipline in that it automatically limits text to the necessary material. In contrast to the conventional suspense outline, the conversational approach holds detail to the minimum consistent with complete reporting.

HOW TO PREPARE THE TOPIC OUTLINE

As demonstrated in the examples that follow, the procedure for preparing the topic outline is short and simple:

(1) Review each statement in the Essential Message in order to decide on a word or phrase that concisely describes the content of that statement.

(2) Write the word or phrase in the margin of the Essential Message, giving it a number or letter as you normally would for an outline (see examples below).

(3) Insert any additional first-order topic headings suggested by the statements.

(4) Break down the first-order topics with second- and third-order subheadings as required.

FOUR EXAMPLES

Example 1—A Technical Report, Before and After

The Situation: Gale N, a drilling engineer in an oil company, conducted a survey of company drilling costs. This entailed visits to all district offices and tedious searches of files and records covering a decade of domestic operations. When he returned to his home office, Gale compiled tables, plotted curves, interpreted data, and concluded that overall costs had declined sharply despite an increase in offshore well costs.

After discussing the results with research department specialists, Gale and his boss decided to open a project to discover new and promising areas of cost-cutting research.

At the time of reporting, Gale was well into a literature search for the new project; he was in no mood to write a report on old work. Without considering his management readers, he hastily prepared a standard outline (Fig. 14) and took it to his boss for approval.

Gale's outline faithfully followed the suspense pattern, from Introduction (a graceful entry with formal statement of objectives and history) to Summary, Conclusions, and Recommendations: what he had set out to do, what he did do, and what he was planning to do. By the time he got to Part V, there were no "Recommendations" to report without repeating, for the only recommendation Gale had to offer (cost-cutting research) was already mentioned (rather, implied) in Part IV-E.

Between his first and last sections there were three typically length-building sections: Methods (*careful* analyses, *comprehensive* surveys, and *detailed* studies—how would you like to read about *careless* analyses for a change?); a Data unit (tables, figures, and discussion of HOW Gale arrived at them); a Discussion section (repetition of data in words, with interpretations and discussions); and, buried in Part IV, the NEWS.

That outline got a response from the boss, but not what Gale wanted. Their conversation went something like this:

Boss: What's new, Gale? What have you got here?

Gale: Well, we've completed this very extensive survey—

Boss: What did you get out of it?

Gale: Oh, the new project. We're planning to—

COST SURVEY

by Gale N.

I. INTRODUCTION

   A. Drilling Costs
   B. History
      1. Cost-cutting research
      2. Value of research?
      3. Question leads to survey

II. SURVEY METHODS

   A. Records — Files — 10 years
   B. Definition of Terms
   C. Cost Breakdown
   D. Compilation and Interpretation of Data

III. DATA (RESULTS)

   A. Comparison
      1. Oil Wells
      2. Gas Wells
      3. Offshore Wells
   B. Comparison by Depth
      1. Shallow Wells
      2. Deep Wells

IV. DISCUSSION

   A. Oil Wells
   B. Gas Wells
   C. Offshore Wells
   D. **Deep vs Shallow Wells**
   E. **All Wells** ⟵   *The NEWS is*
   F. New Project   *buried here!*

V. SUMMARY, CONCLUSIONS, RECOMMENDATIONS

Figure 14   Gale N's original suspense outline.

Boss: But what came out of the survey? Didn't you learn anything of interest?

Gale: Of course! You're interested in profits, aren't you?

Thus Gale was forced to think of his **product** instead of his **labor** as the NEWS of his report. He composed an Essential Message based on his reader's interest, and he used the Essential Message to construct a new Topic Outline, shown in Fig. 15. (Note that the NEWS appears in the first side-heading).

The result was an impressive and *automatic* reduction in report length. Compare the engineered outline of Fig. 15 with the origi-

---

DRILLING COSTS SURVEY

by Gale N.

ESSENTIAL MESSAGE   NEW TOPIC OUTLINE

WHAT is the NEWS?

   Overall drilling costs declined ____ percent in 19____.

WHY?

   Not applicable (not determined).

HOW?

   We obtained this information from a survey of all district records for the last 10 years.

NOW WHAT?

   We have opened a new project to find possibilities for cost-cutting research.

I. Overall drilling costs declined
   A. Oil well costs - down
   B. Gas well costs - up
   C. Offshore well costs - up

II. Survey methods

III. New project started

Figure 15   Gale N's engineered report outline. (See Step 5, p. 99 for an expansion of this outline.)

nal outline of Fig. 14 (and see examples in Step 5, p. 99, for the complete expanded outline).

Example 2—Same Report, Different Reader

The Situation: Assume that Gale N's original report, which was supposed to be written for management interests, had to be written in memorandum-report form for a different audience: the research people who would work with Gale on the new project.

To: RESEARCH GROUP Q

From: Gale N.

Ref.: NEW RESEARCH PROJECT: DRILLING COSTS

ESSENTIAL MESSAGE   TOPIC OUTLINE

WHAT is the NEWS?

   We have opened a new research project to seek ways of reducing drilling costs.

I. New project opened

WHY?

   A survey showinng that drilling costs were reduced _____ per-center in 19____ suggests three areas for investigation.

II. Areas for investigation
A. Oil wells
B. Gas wells
C. Offshore wells

HOW?

   I have started a literature search to . . .
We will then continue with . . .

III. First step Literature survey

IV. Follow-up program

NOW WHAT?

   The assignments for our group are as follows: . . .

V. Group assignments

Figure 16 Organization of Gale N's report to a different audience.

The Analysis for Example 2 produced an entirely different message, even though based on the very same project and information. **A change in readers will nearly always produce a change in the NEWS,** and therefore a change in the **Essential Message,** the **outline,** and the **report.** (Fig. 16 shows how Gale N's report was changed.)

Example 3

An Analyst's Stock Evaluation

This example (Fig. 17) does not involve an intracompany communication, but the readers' primary interest is exactly the same as that of corporate executives: the "bottom line." The securities

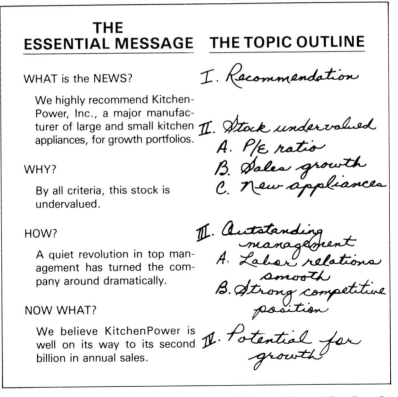

THE ESSENTIAL MESSAGE **THE TOPIC OUTLINE**

WHAT is the NEWS?

We highly recommend Kitchen-Power, Inc., a major manufacturer of large and small kitchen appliances, for growth portfolios.

WHY?

By all criteria, this stock is undervalued.

HOW?

A quiet revolution in top management has turned the company around dramatically.

NOW WHAT?

We believe KitchenPower is well on its way to its second billion in annual sales.

I. Recommendation

II. Stock undervalued
   A. P/E ratio
   B. Sales growth
   C. New appliances

III. Outstanding management
   A. Labor relations smooth
   B. Strong competitive position

IV. Potential for growth

Figure 17   An analyst's stock recommendation to clients. (See Step 5 for complete expansion of this outline.)

THE
ESSENTIAL MESSAGE   THE TOPIC OUTLINE

WHAT is the NEWS?

Now for the first time a relatively simple and economical method of building a hard-surface roadway in the Rub al Khali is available to us.

I. Hard-surface roads for the Rub al Khali

WHY?

This new development is especially important to us in the Rub al Khali, where rugged dessert conditions and remoteness prohibit the use of standard road-building materials.

[included under I, above]

HOW?

(1) The method uses sabkha material usually found within a few kilometers of the roadway site, and the construction work can be done with our own equipment.

(2) We have built and tested some trial sections and found the roadways to be durable and inexpensive.

II. New construction method
A. Material
B. Procedure

III. Tests of the method
A. Trial sections
B. Low-cost construction

NOW WHAT?

In order to get senior management approval for a trial of the method, I propose an early meeting at which we will outline a test program involving two sections in the southern part of the Salwa-Shaybah Trail.

IV. Further tests proposed

Figure 18   Topic outline for a civil engineering report.

analyst who prepared this evaluation knew that his firm's busy clients did not want to have to search around through a badly organized report just to find out if this particular stock suited their portfolios.

Example 4

A Civil Engineering Report

Even a fairly short text can benefit from the detailed, logical organizing possible with Steps 3–5. This example (Fig. 18) is based on a report written by an Aramco engineer in Saudi Arabia.

HERE IS WHERE WE NOW STAND:

Step 4 — Bases the selection and preparation
of illustrations on —

Step 3 — A reader-oriented Topic
Outline, which was based on —

Step 2 — A reader-oriented Essential
Message, which was developed from —

Step 1 — The main point of interest to readers
— the NEWS — determined from the
Analysis of the People Involved.

Step 4
Assemble Illustrations

An ancient Chinese proverb says that one picture is worth "more than ten thousand words." Whether it is one thousand (as the proverb is commonly misquoted) or ten, there is no doubt that the quality of illustrations is every bit as important as the quality of writing. This is particularly true for today's visually oriented generation of readers weaned on television, and it behooves the on-the-job writer to use as much care in selecting and preparing illustrations as in choosing the words of his prose.

At the outset we need to make clear that we are talking about illustrations in the broadest sense, including tables, graphs, charts, line drawings, halftone plates (photographs), and computer-generated figures. We will not cover special types of illustrations such as engineering drawings and geologic maps, which are already part of the specialist's tool kit. Neither do we go into the mechanics of preparing the line drawings, graphs, or halftones—activities that, under the impact of computer graphics, are changing too rapidly to be treated in a book on writing. Rather, this chapter offers tips for making the most of tables and other visual aids to prose description—from the author's viewpoint as well as the reader's.

Since illustrations are usually assembled long before this stage of a writing project, you might well ask the question addressed in the following section.

WHY WAIT UNTIL NOW?

As important as illustrations may be to us, with our total project reflected in tables, figures, etc., we don't want them to lead us around by the nose. In particular, it's a big mistake to let tables and figures determine the structure of the document.

When on-the-job writers finish a project, they typically have an impressive stack of illustrations at hand. Many of these, having been accumulated from previous interim reports, will be arranged in the chronological order in which the data were acquired—a naturally suspenseful process. Perhaps the very first classic suspense report ever written was produced by some historic scientist who, after looking over his bountiful collection of tables, graphs, and drawings, decided to make his text conform.

This is an ego-inflating arrangement, for describing the lot ensures that no whit of the writer's effort will be denied posterity. Unfortunately, the chronological arrangement also builds the longest possible report, delays and buries main points, and risks overwhelming readers with more than they want to know.

If you look at your own pile of illustrations and start thinking in that traditional pattern, you may be tempted to revise the Topic Outline developed in Step 3. Don't let that happen. Your new outline is based on a series of steps that reliably defined the readers' needs and your own objectives. The framework of your document is already established to meet them satisfactorily. Your task now is a simple one: you have only to review each topic in your outline, select or prepare the supporting illustrations, and assemble them in the order of outline topics.

Although using the Topic Outline to guide the sequence of illustrations does not answer all questions about the need for data in tables and figures, it does ensure that the data selected will be what readers need. For instance, if you were writing a report to a construction engineer, the Analysis of People might have produced a Topic Outline allowing for some sizeable tables. One might provide a detailed cost breakdown. How much for materials (for 5/8″ pipe? for 2″ pipe? etc.). And, similarly, how much for each piece of equipment or spare part, and how much for each type of labor?

But for a report to that engineer's vice-president, the analysis would have led to an Essential Message including only a management-type table—that is, a summary showing just the subtotals for materials, equipment, and labor. This sort of table could be included within the text.

If the readers included both the V.P. *and* the construction engineer, your Analysis of People would have produced a Topic Outline placing data summaries up front, with increasing detail in subsequent sections and great detail in appendices. Readers could drop out at will, being neither deprived of nor overburdened with detailed information.

At this stage you should have your illustrations in roughly the proper sequence, but you may still have more or fewer of them than you need to support your conclusions and interpretations. This brings up another question.

HOW MUCH DATA SHOULD BE INCLUDED IN ILLUSTRATIONS?

Getting to the point at each major level of a report requires that illustrations be brought in to back up, rather than to introduce or lead into, items of NEWS (conclusions, results, interpretations, etc.). For that reason, it may turn out that the last figure generated should be the first to be displayed. The end-product illustration frequently proves to be the most compelling support for an idea. In this case, illustrations generated during intermediate steps in a project are included only as needed to explain how the end product was arrived at—or perhaps are put in an appendix to satisfy a few readers interested in that much detail.

Great masses of raw data are usually of little interest to most targeted readers, but leaving out hard-won raw data can involve difficult decisions. As authors, we tend to believe that someone out there will surely need it all. Occasionally that is true, especially in research organizations. If so, there are ways to make volumes of raw data available without turning the document into a dissertation of formidable bulk.

For example, you can write a supplementary report (giving it a reference number related to that of your report for distribution), stash the supplement with its treasurehouse of raw data in the files, and include in the front of the main report a tear-off coupon for ordering the supplement. You may find, as most organizations do, that takers are few and far between. But those few could be the most important people in your technical audience. For instance, access to your work might stop future researchers from repeating work already done.

In any case, the objective is to provide the backup needed without either overwhelming or shortchanging the reader. Try limiting your first set of illustrations to those prompted by topics in the outline and statements in the Essential Message. When you have expanded the outline (Step 5), the collection of illustrations can be reviewed. You may need more or fewer, but you will end up with a reader-oriented selection rather than a pack rat's collection.

WHAT ARE THE BEST
WAYS TO DISPLAY DATA?

Computer software is giving the above question new relevance. Now, with increasing facility to choose from a variety of ways to display the same data, we have to think about what impression we wish to leave with the reader. In some situations a table will serve best; in others, a graph or chart may be preferred for its visual impact. The choice also depends, of course, on the type and amount of data, space limitations, company guidelines, and budget restrictions.

Readers' technical limitations may also be a factor. For example, some data may be simpler for you to plot on log-log or polar-coordinate graph paper and may be more elegantly displayed that way; but if your readers are unfamiliar with logarithms or polar coordinates and would be confused by their use, give up the idea or else devote a part of your text to explaining them.

Table 1 lists advantages and disadvantages of the three modes and five main types of data presentation. All can be handled by computer, if we include simulated photographs (which can be surprisingly good). Visual aids for talks, a special case, are covered in Part V (p. 252).

Tips on Preparing Illustrations

(1) Keep illustrations lean and clean.

A cluttered illustration affects readers in the same way a solid page of fine print does. If a graph has too many overlapping curves shown by different symbols (e.g., ..., – – – – – – – – –, + + + + + + + + , = = = = = = = = = = =), simplify it. Either separate the information so it can be shown on two or more matched figures, all at the same scale, or (if the budget permits) use color.

The size selected for illustrations may be a compromise involving readability, convenience, and cost. If space is not a problem, a small figure that looks cluttered can be uncluttered without deleting anything by enlarging it several times and then redrafting the letters and symbols at something close to their original size. The net result: more white space.

(2) In the text accompanying illustrations, state the conclusion (make the point) that you want readers to grasp.

This extends a courtesy and a convenience to readers and serves your purpose of getting maximum impact from the illustrations. Many writers, for example, would lose this benefit by introducing the illustration in Fig. 19 like this:

Fig. 19 shows the annual average value of the dollar against 10 foreign currencies during the period 1975 to 1983.

A glance at the caption confirms that this is a statement of the

TABLE 1 Types of Data Presentation: A Generalized Comparison

Mode	Type (in black-and-white or color)	Advantages	Disadvantages	Cost
Verbal	Descriptive Text	Flexible; good for highly generalized descriptions, conclusion, or summaries.	Minimum visual impact. Unsuited to describing complex numerical or physical relationships in detail.	Lowest
Numerical	Tables	Two-dimensional representation of verbal information (as in this table) or digital data. The only practical way to present large amounts of detailed numerical information, e.g., financial reports.	Less visual impact than the types shown below. With vast quantities of data (e.g., giant spreadsheets), tables can become so complex and unwieldy that readers are discouraged from studying them.	generally increasing costs due to labor (time), materials, and reproduction
	Graphs: bar charts, pie diagrams, x-y curves, polar coordinate graphs, isometric (x-y-z) graphs, etc.	Good for presenting digital (numerical) data in analog format, thus providing an overall view of a relationship that can be grasped at a glance.	Precise values are not emphasized. Bar charts and pie diagrams are not well suited to showing complex relationships. Nonlinear scales have to be explained to people unfamiliar with them. Highly complicated graphs can be hard to understand without a lot of study.	
Visual (Pictorial)	Line Drawings, Computer Drawings	Indispensable for the presentation of certain kinds of information (plans for construction; engineering designs; maps; etc.). "One picture is worth more than 10,000 words."	Complex drawings can be expensive, although facilitated by computer. Compared to a photograph, a drawing is one step removed from reality.	
	Halftone Plates, Color Photographs	Visual interest and authenticity: "Seeing is believing."	Higher cost. Unless photos are particularly clear, readers may not see what they are supposed to see.	Highest

Figure 19   How would you introduce this figure?

obvious. The author should draw attention to the point made by
the figure:

> As Fig. 19 shows, the dollar actually remained below the
> March 1973 baseline for nearly three years running, from
> early 1978 to late 1980.

(3) Locate captions consistently.

The standard for figures is at the bottom; for tables, at the top.
If a halftone or color photo takes up an entire page, its plate number and title should be put either on the back of it or on the facing
page. Very large illustrations such as maps may have the caption
within the border rather than underneath it.

(4) Facilitate comparisons in tables.

Here is a good way to prepare effective tables of comparison:

(a) Begin by stating the problem of comparison in words that show (1) what you are ultimately comparing and (2) what characteristics, properties, or other attributes you will be using to make the comparison (see Tables 2 and 3).

(b) Arrange items for comparison (especially digits) in vertical columns; readers find it harder to compare items in a horizontal row.

(c) If you are dealing in days, months, or years, readers normally expect time to run from left to right (this of course applies to graphs as well as tables). Only geologists think of (geologic) time as running from the bottom to the top of the page.

(d) Use the **stub** (left-hand column) of the table to list the things that are being compared; use the column headings to name the characteristics used in the comparison (Tables 2 and 3).

(e) Make sure the headings (including units of measurement) correctly identify data in columns beneath them.

(5) For management tables, focus on highlights.

For memos or reports to management, use short tables illustrating a single concept. Such tables are often introduced by a sentence ending with "as follows:" or a similar phrase and are inserted right into the text.

(6) If a table gets too complex or too big to handle, prefer (in the following order) to:

(a) Break it up logically into self-contained subtables.

(b) Turn it broadside on the page.

(c) Have it typed on a large sheet and reduced photographically to page size.

(d) Use a gatefold (a fold-out sheet).

TABLE 2
Preparing Tables of Comparison

A. A poor statement of the problem:

To compare the **porosities, permeabilities,** and **grain sizes** of rock samples L, M, and N.

—And a poor table arising from that description of the problem:

| | Rocks | | |
Characteristics	L	M	N
Porosity, %	2.0	8.5	16.0
Permeability, md	0.8	13.0	105.0
Average grain size, mm	0.02	0.09	0.6

B. An accurate statement of the problem:

To compare **rocks** L, M, and N with respect to three of their characteristics: porosity, permeability, and average grain size.

—And a good table based on that statement*:

| | Characteristics | | |
Rocks	Porosity, %	Permeability, md	Average Grain Size, mm
L	2.0	0.8	0.02
M	8.5	13.0	0.09
N	16.0	105.0	0.60

*Note that Table B meets criteria for clarity: the elements to be compared appear in the stub; the characteristics by which they are compared appear as headings over columns; headings and units of measurement accurately describe what appears in each column; and values to be compared are aligned vertically.

TABLE 3
Example of the Right Way to Set Up a More Involved Comparison

Statement of the problem:

To compare **three methods of analysis** (I, II, and III) as performed by laboratories E, F, and G, with a standard and with each other, *with respect to* their accuracies in determining concentrations of carcinogen V in the tap water and groundwater from three different cities (P, Q, and R).

The table derived from this statement:

| Method | Concentration of V, parts per billion | | | | | |
| | *City P* | | *City Q* | | *City R* | |
	Tap	Ground	Tap	Ground	Tap	Ground
Standard	9	11	2	4	15	36
Method I						
Lab E	7	12	—	3	21	
Lab F	8	11	—	5		
Lab G	8	13	1			
Method II						
Lab E	14	14				
Lab F	11	14				
Lab G	12					
Method III						
Lab E						
Lab F						
Lab G						

(7) Consider putting large tabular summaries of record in an appendix.

Some of the new accounting software for computers will generate spreadsheets that are absolutely gigantic. But even when treated as suggested in the preceding entry, they can still make pretty heavy going for the reader. Shorter tables illustrating specific points are more suitable for the text.

(8) Place page-size or smaller illustrations as close as possible to the sentence that introduces them.

Prefer first to place illustrations immediately after the sentence; second, on the facing page; and last, on the following page.

(9) If the reader must refer frequently to a lot of illustrations, attach them to a foldout flap at the end of the report.

Where text tends to get lost in a profusion of illustrations, putting all the figures together at the end of the text helps somewhat, but it still leaves the reader to do a great deal of flipping back and forth. One effective but more expensive solution to the problem involves three bindings instead of one; it is shown in Fig. 20. The right-hand flap holds the illustrations, bound at the top, so that the reader can page independently through text and figures without losing his place.

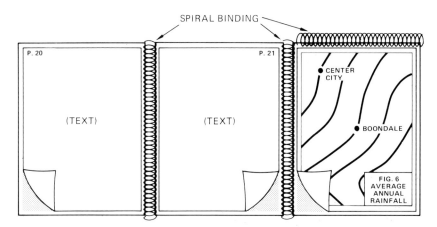

Figure 20   Solving the problem of too many illustrations, little text.

(10) Strive for impeccable work.

It is a fact of psychology: neatness adds credibility. A carefully drawn diagram looks more correct than a rough sketch; a professionally drafted and colored map, or a computer-generated graph, appears more reliable than one hand-drawn in pencil.

THE NEXT
STEP

Your selection of illustrations for a long report will be further refined—perhaps reduced or increased—by the next step: preparing an Expanded Outline. When each major topic in the outline has a sentence or two stating its main point, you have a two-way advantage: the sentences suggest illustrations, and the illustrations suggest the most effective statement of the main points.

If you are working on a short document, you are now ready to proceed to Step 7.

The expanded outline is not this difficult if a few precautions are observed.

Step 5
Expand the Topic Outline

Expanding a Topic Outline extends the process of organizing to enable you to **get to the main point at the beginning of every major section in a report.**

The procedure is simple: just open up the Topic Outline and insert, at each major level, a Summarizing Topic Sentence. As explained below, a Summarizing Topic Sentence not only introduces the subject of a unit; it also makes the point of the unit.

This Expanded Outline serves four important purposes:

(1) It produces a report text that can be scanned quickly for the overall "story." Lead sentences highlight the important points throughout the report.

(2) It reduces document length automatically and effortlessly— far more than any amount of editorial red-penciling could do.

(3) It gives the writer ready-made starting sentences for each topic, which facilitates skillful dictation (getting started is often the hardest part of writing).

(4) Finally, with illustrations attached, the Expanded Outline forms, in words and pictures, all you need to call for a Prewriting Conference. As explained in Step 6, this conference allows all reviewers, including those who might modify or reject your report, to provide their input and approval **before you write.**

USE SUMMARIZING TOPIC SENTENCES

In high-school English courses we learned to write "topic sentences" (or lead sentences), which introduce a topic but rarely do much more than identify it. They do not necessarily make any point at all. Summarizing Topic Sentences, on the other hand, do both.

The difference between the two sentence types is illustrated in the following examples ("theme-setting" sentence skeletons are underlined):

Simple Topic Sentence	**Summarizing Topic Sentence**
<u>Efforts</u> to improve the exhaust valve's action by making the valves out of plastic instead of metal <u>continued</u> during the past month.	Continued <u>testing</u> during the past month <u>shows</u> that plastic <u>valves</u> <u>are</u> no <u>better</u> than metal ones.
(This leads into a description of "efforts.")	(This leads into WHY they are no better and then HOW the tests proved it.)
Mr. Mannijer <u>disscussed</u> our <u>continuing</u> <u>interest</u> in the level of accident rates last year.	Mr. Mannijer <u>reported</u> that last year's accident <u>rates</u> <u>increased</u> in all departments as follows:
(This leads into a philosophical discussion of corporate objectives.)	(This leads, obviously, into hard NEWS.)
A <u>study</u> of the feasibility of <u>moving</u> our central plant from its present location in downtown Biggs City to the Exville locale <u>was carried out</u> using the following criteria as suggested in our April 20 meeting.	A feasibility <u>study</u> <u>shows</u> that the Exville <u>location</u> <u>meets</u> all our <u>criteria</u> for relocating the Biggs City plant, as follows:
(Now we get a repetition of criteria laid out in the April 20 meeting.)	(Now we learn HOW Exville meets the criteria.)
<u>Tests</u> <u>were conducted</u> to determine whether the special <u>alloy</u> developed by Miracle <u>Metals</u> Inc. could reduce the <u>weight</u> of <u>wide-body</u> passenger <u>aircraft</u>.	<u>Tests</u> <u>have demonstrated</u> that a special <u>alloy</u> developed by Miracle <u>Metals</u> Inc. <u>can cut</u> the <u>weight</u> of wide-body passenger aircraft by close to 15 percent.
(This is a classic introduction to a METHODS section; typically it leads into a description of the tests, with results delayed.)	(This gets to the point of the tests described in the METHODS section.)

We might view starting sentences like those in the left-hand column as the beginning of a mini-suspense format, patterned after the traditional suspense format for an entire report. Thus the whole text would be, in effect, a series of small suspenseful reports within a suspenseful report. And that is what increases length unnecessarily.

The shorter and more scannable a report is, the more likely it is that people at upper levels of the hierarchy will read it, or at least scan it. Too often, though, from your viewpoint as a writer, managers pass long reports on to a middleman—an assistant who is obliged to study the report and highlight the key points not included in your summary.

The middleman may do a fine job. On the other hand, if he has to compose the highlights from scratch, he may provide a quick, skimpy summary that does not do justice to your work. Moreover, lacking clear-cut interpretations (or conclusions or findings) in the major units, he may justifiably report his own. Such third-party interpretations have been known to actually contradict what the author thought he was conveying by presentation and discussion of data.

If you make your own main points at the beginning of each major section, they can be plucked out easily—in *your* words.

EXAMPLES OF EXPANDED TOPIC OUTLINES

Example 1—Gale N's Report on Survey of Drilling Costs

The Situation: Refer to Step 3, p. 76 for the background on this project. Gale's revised Topic Outline (Fig. 15) was expanded with a statement of overall NEWS at the beginning and a main point for every major unit, as follows:

I. **Overall Drilling Costs Decline**
   The Company's overall drilling costs declined in 19__, despite a rise in offshore well costs.
   A. Oil Well Costs—Down
      Oil well costs decreased _____% between 19__ and 19 __, from $__ to $__ per foot.
   B. Gas Well Costs—Down
      Like oil well costs, gas well costs declined steadily, from $__ per foot in 19__ to $__ this last year.
   C. Offshore Well Costs—Up
      The only costs that rose during the last ten years were those for offshore wells.

   D. Shallow vs Deep Wells
      Throughout the ten-year period under study, the per-foot
      costs of shallow wells were lower than those for deep
      wells.
 II. **Survey Methods**
      These figures were obtained in a survey of all company dis-
      tricts over the past decade, along with a literature survey of
      general industry experience.
III. **New Project Started**
      In view of the information turned up by the survey, we have
      opened an exploratory project with the object of finding prom-
      ising new areas of cost-cutting research.

   To appreciate how Gale got to the point at every major unit
within his report, look at Fig. 21, a graphic representation of his
first chapter, "Overall Drilling Costs Decline." A main point begins each of the four units in the chapter.

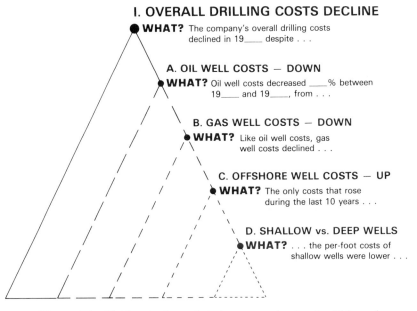

Figure 21   Getting to the point at every major level within a document.

Example 2—An Analyst's Stock Recommendation

   Refer to Step 3, Fig. 17 (p. 80), for the Essential Message and
Topic Outline that produced this Expanded Outline.

I. **Recommendation**

   We strongly recommend KitchenPower Inc., a major manu-facturer of large and small kitchen appliances, for growth portfolios. It is currently trading around 28.

II. **Stock Undervalued**

   By all criteria, this stock is undervalued.

   A. P/E Ratio

   Its recent P/E ratio of 5.2 is near the bottom of its multiple range over the past four years (4.5 to 18).

   B. Sales Growth

   Sales have nearly doubled in each of the last three years.

   C. New Appliances

   The company is preparing to announce a hot new line of computerized appliances.

III. **Outstanding Management**

   A quiet revolution in top management has turned the company around dramatically.

   A. Labor Relations

   Union relations are excellent.

   B. Competitive Position

   The company is competing very effectively with European and Japanese manufacturers, and the new computerized line should give it a decisive advantage in the North American market.

IV. **Potential for growth**

   We believe KitchenPower is well on its way to its second billion in sales. The strength for continued growth is there.

Note: In both the above examples the writer has the starting sentence for each unit of the report. From those starting points he can dictate easily, expanding, elaborating, explaining, giving examples, etc., as needed under each starting sentence. Fig. 22 is a condensed version of the final ''Investors' Review'' dictated from the expanded outline of Example 2.

Example 3—A Formal Proposal to a Foreign Government (Appendix B)

The example included in Appendix B, based on the most complex reporting situation in our files, demonstrates the applicability of Engineered Writing to very difficult projects. This project is fraught with problems—political, personality, technical, and cultural—and involves millions of dollars. The example begins with a description of the situation and continues from Step 1 through Step 5.

INVESTORS' REVIEW

STOCK NAME	KitchenPower Inc.
RECOMMENDATION	We strongly recommend Kitchen-Power Inc., a major manufacturer of large and small kitchen appliances, for growth portfolios. It is currently trading around 28. During the last 12 months it has ranged from 25 to 41. (The all-time low of 19 was recorded back in 1987.)
STOCK UNDERVALUED	By all criteria, this stock is undervalued.
— P/E Ratio	Its recent P/E ratio of 5.2 is near the bottom of its multiple range over the past four years (4.5 to 18). It is also below the P/E ratios of any of its competitors.
— Sales Growth	Sales have nearly doubled in each of the last three years. Overseas sales are particularly strong, thanks to shrewd use of native sales personnel in more than 30 countries.
— New Appliances	The company is preparing to announce a hot new line of computerized appliances that will put it at least one year ahead of the competition. The new line is expected to be particularly well received by consumers in North America, as well as in Europe and Japan.
OUTSTANDING MANAGEMENT	A quiet revolution in top management has turned the company around dramatically.
— Labor Relations	Union relations are excellent. They began to improve after two moderate union leaders were named to the board of directors and are now considered the best in the industry.

— Competitive Position	The company is competing very effectively with European and Japanese manufacturers; it has maintained an average 17% market share in Western Europe and 12% in Japan. The new computerized line should give it a decisive advantage in the North American market, at least for another year.
POTENTIAL FOR GROWTH	We believe KitchenPower is well on its way to its second billion in sales. The strength for continued growth is there, particularly in view of its promising early steps in the potentially huge Chinese market.

Figure 22 Text of a stock recommendation dictated from the expanded outline.

Example 4—A Memo Based on an Expanded Outline

The following two versions of a memo written to recommend the purchase of a camera demonstrate the difference between a well-engineered text and the standard historical document. Version 1 is based on Steps 1, 2, 3, and 5. (This is a case in which it was best to combine most of the WHY answers with HOW answers.) Version 2 of the same memo is about 40 percent longer, and there is no highlighting of main points. Which version would you rather receive?

Version 1—A Well-Engineered Memo

To: L. Heffay   Date: 27 February 1989
From: P. Ohn   Ref: New Camera for Microfiche

What? I recommend leasing a Brand M step-and-repeat camera at $300 per month for use in the production of the microfiche edition of our company reports.

why? This equipment will replace the rented planetary camera and the microjacket filler now in use.

why and The step-and-repeat camera will:

- Reduce costs by about $1,200 per year.

How? Production cost per image is $0.006 lower—a saving of $1,200/year at our present volume of 200,000 images/year. The step-and-repeat camera eliminates manual operations. It records the original images directly onto 105mm (4″ × 6″) film, producing a finished microfiche in a single step. Our present equipment requires that the images be recorded on roll film, which is then cut into strips and inserted into micro-thin jackets. Contact prints are made from these jackets to produce the finished microfiche.

- Increase image quality by 50 percent.
This new camera has better resolution and produces microfiche with first-generation images. The present microfiche contain third-generation images, and quality is further reduced by the filter effect of the jacket.

- Reduce file size by about one-third.
Operating at a 42X reduction ratio, we can produce a 4″ × 6″ microfiche containing 208 images instead of the present 60 images.

- Provide compatibility with COM-generated microfiche.
Standardization of microforms would permit standardization of reading equipment.

Attached for your approval is a requisition for the leasing of a Brand M step-and-repeat camera to produce our reports in microfiche form.

(224 words)

Version 2—Original Suspenseful Organization

To: L. Heffay   Date: 8 February 1989
From: P. Ohn   Ref: Company Reports in Microform

In 1972 we began publishing our company reports in microform and thus made considerable savings in printing and distribution costs. The process for making these microforms (4″ × 6″ microfiche) is, however, rather cumbersome and could be improved.

Under the present system, the reports are microfilmed using a 16mm planetary camera. The film produced in this operation is processed and cut into strips about 6″ long and inserted into microjackets (a clear acetate sheet 4″ × 6″ with five chambers or pockets into which the microfilm strips are fed). The microfiche are produced by making a copy of the jacket onto a photographically sensitive sheet of acetate film.

As you can see, this process requires several manual steps to transfer the original microfilm image to the form in which it is finally used. Microfilm cameras that can record the original image on sheet film and thus eliminate the intermediate steps have been available for some time, but their cost has been very high.

Now, however, several manufactures offer these step-and-repeat cameras at prices that make them attractive for our use. One in particular can be rented for about $300 per month.

The acquisition of this camera would enable us to produce microfiche for about $0.019 per image as compared to $0.025 for each image by the present method. This would result in a saving of $1,200/year at our present volume of 200,000 images per year. And the quality of the microforms would improve about 50 percent on account of the improved resolving power of the optics in the step-and-repeat camera.

In addition, the microfiche produced would be identical in format with those produced in the Computer Output Microfilm (COM) operation now used for our statistical reports. As the COM format provides 208 images per 4″ × 6″ fiche as compared to 60 images per fiche in the jacketing method, the volume of our microfiche file would be reduced threefold. Standardization of microfiche reading equipment also becomes a possibility with standardization of the microform.

WHAT? (The NEWS)

7 For these reasons, it is recommended that we lease a Brand M step-and-repeat camera to replace the present planetary camera and microjacket filler now in use to prepare our reports in microform. I will be happy to provide any additional information you may need to evaluate this proposal.

(377 words)

OTHER USES FOR THE EXPANDED OUTLINE

The Expanded Outline is very useful to speechmakers (see Part V, p. 252). It can also be put to good and continuing use in setting up formal plans (writing planning reports) for new projects. The outline is developed, as usual, from an Essential Message based on an analysis of the people who need, have requested, or would benefit from the proposed work.

In general, the Analysis of People will answer questions like these:

(1) WHAT (end product, information, outcome) do we WANT? Exactly what is our objective?

(2) WHAT are users' interests? Who needs it? Who asked for it? Who will benefit, and how?

(3) Are there any special (politics, personality, budget, etc.)

problems in setting up the report plan? How will we handle them?

The Essential Message on which the Expanded Outline is based answers questions like these:

WHAT (do we expect to achieve)?
(**Note:** If the objective is successfully attained, this will be the NEWS of the completion report. Indeed, unless special problems develop, it will be the NEWS even if the objective is not achieved.)

WHY (seek this objective)?
Will it solve a problem? Have new or useful applications? Save time or labor? Improve efficiency? Make a profit? Serve as a basis for research (or further research)? Increase understanding of a concept? Extend scientific knowledge?

HOW (No. 1)?
How will the work be carried out? (step-by-step plan)

HOW (No. 2)?
How much time do we have for the project? How many people will be assigned to it? Who will do what?

HOW (No. 3)?
How much money should be budgeted?

NOW WHAT?
What's the next step? Prepare an AFE (authority for expenditure request)? Get higher approval? Buy equipment? Find a site? Hire specialists? Call a meeting? Is timing critical?

The Expanded Outline has been used to organize formal reports describing planned projects (reports for the record or reports to accompany proposals). But its utility extends beyond that. First, when the Expanded Outline is discussed in a Pre-writing Conference (Step 6), input from all parties concerned ensures a well-thought-out plan. Any changes agreed on in meetings are incorporated into the plan in writing, which prevents later misunderstandings. Once the new project is under way, the Expanded Outline serves as an excellent checklist to keep the work on track.

And finally, when the time comes to write a completion report, the first step is to review the plan outline. During the weeks, months, or years between initiation and completion, writers tend to forget why the project was started and who needed the results.

A research biochemist described the problem this way: "I've worked on this project for three years, and the trivia cluttering up my recollection is mind-boggling! But when I went back and consulted my plan outline, a well-turned statement of the NEWS just formed itself!"

Writers often go through the whole chore of organizing, writing, and editing a difficult report, only to have it rejected because the real point is missing. For example, as one writer in a workshop struggled to come up with a main point for a completion report, his manager, monitoring the class, interrupted: "Remember why we started this project? Don't you recall that Jack over in the Southern District brought up a question that we had to find an answer to? *Where's your answer?*"

FINAL THOUGHTS ON
THE EXPANDED OUTLINE

Don't carry the expansion too far. You don't want to end up with a 40-page outline for a 45-page report.

Do make sure your illustrations are properly tied in to the outline's text.

As you go over the rough draft of your Expanded Outline, keep in mind how it will be used at the Pre-writing Conference. Make sure your summarizing topic sentences emphasize what you found out, not what you did.

Step 6
Arrange a Pre-writing Conference

> Around this place there never seems to be time to do it right the first time—but somehow, there's always time to do it over again.
> —Classic Complaint

The Pre-writing Conference is a meeting in which the Expanded Topic Outline is discussed, perhaps modified, and approved before you write the report. Send a copy of the Expanded Outline, with illustrations attached, to everyone who will review and approve (or disapprove) your report, giving them a week or two for study. You can't always get everybody to attend the conference—some will be out of town or in other meetings—but go as high in the hierarchy as possible. If necessary, wait for some of the participants.

BENEFITS OF THE CONFERENCE

These conferences are of demonstrated value to both writers and managers in terms of time, money, effort, internal communication, and morale. The president of an oil company has said that a two-hour report-planning conference with five people is worth all the time and effort expended. If the document involves problems requiring a much longer conference with more people, he said, the benefits will be even greater.

It Prevents Rewriting and Recycling

The advance conference often reveals conflicts that the writer could not be aware of. Even the readership may be redefined; and in that case the NEWS will usually change, as will the amount and kind of supporting detail. There may be technical disagreements. Institutional policies, corporate objectives, national politics—all have been known to complicate what appeared to be a straightforward reporting job.

If these matters get attention only after a report is written, the review, discussion, changes, and rewriting will be more complicated and time-consuming for everybody. After the document is rewritten, the whole review process must begin all over again. The author will have tried to incorporate all the comments and suggestions given piecemeal as the original version made its way up through the approval levels, and the old maxim "a camel is a horse that was put together by a committee" may apply to the final patchwork version.

It Helps Solve Special Problems

If you have had special problems in organizing a document (see Step 1, p. 37), the Pre-writing Conference provides an opportunity to get (1) ideas on solving the problems and (2) approval of the solutions before the writing is set in concrete. However, don't go into the conference unprepared, merely asking for help; the problem should have been analyzed and worked out to the best of your ability in Step 1, and certainly before the Expanded Outline was prepared.

The handling of bad news is an example. Managers and supervisors dislike surprises, especially unpleasant ones. If you have bad news to report, a Pre-writing Conference may provide a way to help solve this often difficult problem. Instead of trying to hide the bad news or write about it in such a way as to lessen its impact, you can report on it orally at the conference and enlist managers' help in wording the news for publication.

It Boosts Morale

Nothing is more frustrating for a writer than having to rewrite a long report because he was not informed of special circumstances, problems, and management requirements—not to mention the fact that somebody up the line had a special reason for including unanticipated readers.

On the other hand, supervisors and managers are equally frustrated by having to spend time in endless conferences to get an adequate report out. Most managers do not enjoy criticizing. Nor do they appreciate having to organize somebody else's writing, especially if that involves reading 30 or 40 pages of text and figuring out what is wrong with it.

Since the Expanded Outline displays both format and technical conclusions and results, the Pre-writing Conference gives everyone a chance to make changes before time, effort, and money are poured into the writing project. You discuss your elaboration; explain key points; answer questions; defend your views (and win or lose); and revise the outline as needed before leaving the conference, even changing the NEWS, the conclusions, and the recommendations if necessary. All this before anybody has had to write or read a lengthy text!

A boost in morale is not merely to be anticipated; it has been demonstrated in many companies where the Pre-writing Conference is standard procedure.

It Enables You to Dictate

Successful dictation (Step 7) is utterly impossible without prior organization of ideas. Moreover, despite all the recognized benefits of dictation, it will be wasted time and effort if the results are not approved.

After a Pre-writing Conference, the writer can compose confidently, using the revised Expanded Outline that has received the approval of all concerned. That confidence makes for an extraordinary difference in tone, style, and even content of a document.

It Promotes Vertical Communication

An unexpected benefit of the Pre-writing Conference is its contribution to vertical communication within the organization. Professional staffers learn things about corporate policy that are not written down in any policy guide. Managers learn things from the technical and administrative staff that nobody would otherwise volunteer. Prejudices, biases, and idiosyncrasies are revealed. People understand each other better, and that promotes good writing as well as good work.

YOUR PART IN THE PRE-WRITING CONFERENCE

Don't take a passive role in the Pre-writing Conference. Be prepared to defend your points of view technically. Ask questions. Be sure your illustrations are approved and your outline revisions are okayed at every step, and summarize the consensus before you leave. No doubt you are aware that what ''they'' think they agreed on is not always what you or others think you heard.

THIS...

...NOT THIS

Step 7
Dictate

It bears repeating: by all means, dictate! If you have not already developed this skill, it is in your professional interest to do so as soon as possible.

WHY DICTATE?

- Dictation is a clear-writing technique.
- It is a speed-writing technique.
- It promotes objectivity—for effective editing.
- It is a cost-cutting measure.
- And it may soon be a required skill.

A Clear-Writing Technique

Dictation gets the natural sound of the human voice—your own, happily—into the prose. That's good, for strong reasons besides the fresh individuality it imparts to your writing. You are entitled to that personal style so long as your document is technically accurate and complete, logically organized, and grammatically acceptable.

Material composed by hand is too often stilted and pompous. It is full of ornate language and passive verbs, which become distorted verbs (see Part III), which attract all sorts of padding words and "impressive" phrasing. We don't talk that way and we shouldn't write that way. In speech (and dictation) we use more active verbs, fewer complex sentences, and simpler language. Our sentence skeletons are meaningful and they establish the right theme: what we intend to talk about.

Rudy Fascell, an executive with the Agency for International Development, put the matter this way: "It's my observation that when people set pen to paper, their thoughts become stylized. They tend to forget about plain speech and plain writing; they feel obliged to write in a formal, stilted manner. Their self-image gets in the way, causing them to come across as very different from the way they really are."

When we write in longhand, our attention centers mechanically on the words and sentences we see before us. Ideas get short shift in the attempt to perfect each little part of a sentence and each

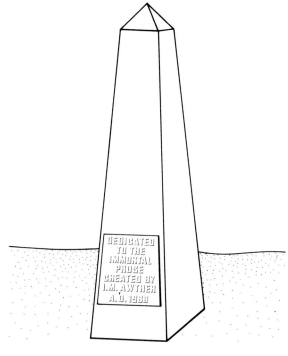

Monumental prose

individual sentence within a section. Talking, however, is a natural, practiced act, so we are able to dictate while concentrating on ideas, which is what we should be doing at this stage of the project.

Talking a document through gives it a relaxed, conversational tone that bespeaks confidence and dignity—even formality when desirable—without stiffness. Learning to dictate easily may take a few trials, but minor learning problems should not be used as an excuse to give up dictation. With practice you can dictate hard-hitting sentences that demand action, in a fresh style that is uniquely your own. The incentives for acquiring this skill are compelling.

A Speed-Writing Technique

Although your brain generates ideas more rapidly than you can state them, you can speak them at about 150 words per minute—

four to six times as fast as you can write them in longhand. This will give you some idea of the time savings available through dictation.

To further illustrate the value of dictation in terms of time, we recall the experience of a research chemist who decided to test dictation vs handwriting. She had just finished the first of two similar service reports (reports about work done for and charged to affiliated companies). The subject of the first report was application of a research-developed technique to a problem experienced by an affiliate in lab work. The second report was similar, but it applied a modification of the technique to a different problem in a different lab. Both reports were necessarily lengthy and complex.

Here is a record of the time the chemist spent on her first report, which was written in longhand:

Writing, revising, and perfecting first draft	60.5 days
Conferences in which different versions were rejected by different supervisors	3.5
Rewriting after successive rejections	39.0
Editing the final draft	5.2
Total time	108.2 days

It was this miserable 3½ months that prompted the chemist to give dictation a trial. In preparing her second report, she presented an Expanded Outline in a Pre-writing Conference attended by everyone who could modify or disapprove the report. After a few revisions had been made during the conference, the content and format were approved, and off the chemist went to find a dictating machine. Then she sat down and used the Expanded Outline, with its convenient starting sentences throughout, to dictate the report. Here is the time record on which charges to the affiliate company were based:

Preparation of Expanded Outline (Steps 1–6) (She had help from an editor here, for it was her first experience with the system)	1.5 days
Attending Pre-writing Conference	0.5
Dictating the text	1.0
Editing the final draft	1.0
Total time	4.0 days

The most interesting part of this story is that the chemist's first report was still waiting its turn in the Word Processing Group, and she was able to recall it and bring its quality up to that of her second report.

A Means of Promoting Objectivity and Effective Editing

Words written in longhand take up permanent residence in the mind. Anyone who labors to write in longhand makes a big investment. Some people even think that suffering is an essential part of writing. It's a time of searching, groping for the *mot juste*—the supremely precise expression. Sentences are revised, polished, and repolished. Entire sections are consigned to the wastebasket and rewritten. The search for perfection leaves sweatmarks, and maybe figurative drops of blood, all over the final version.

Worse yet, in reaching "perfection" the writer develops a love affair with his prose. He is blind to its faults. If a supervisor makes or requests changes, he might as well be saying that the writer's child is homely; criticism hurts.

Dictated material, on the other hand, leaves the mind too quickly to become cherished. The document comes back from the typist as a stranger's prose. Editing is fast, easy, and painless. Flaws can be recognized and corrected before someone else catches them.

A Means of Increasing Productivity

Studs Terkel tells in his book *Working** of an interview with the president of a radio and TV corporation, who commented on a typical morning's work:

> "My day starts between four thirty and five in the morning, at home in Winnetka. I dictate in my library until about seven thirty. Then I have breakfast. The driver gets there about eight o'clock and oftentimes I continue dictating in the car on the way to the office. . . . I talk into a dictaphone. I will probably have as many as 150 letters dictated by seventhirty in the morning . . . ."

---

*Studs Terkel, *Working*, (New York: Pantheon Books, 1972), p. 390.

This is not to suggest that you should start your working day in the wee hours before breakfast or that you should aim for 150 letters by 7:30 a.m. But this executive's accomplishment does demonstrate how dictation can increase productivity.

Dictation is an excellent means of making otherwise wasted time profitable. If you commute daily or travel a lot, a portable recorder will enable you to dictate summaries of meetings, conversations with clients, proposals, invoices, promised letters, etc., while you spend tedious hours on trains or planes, or in airports. If a job takes you to the field frequently, you can dictate reports conveniently while your ideas are fresh—walking, driving a field vehicle, or bouncing about in a helicopter or light aircraft. You won't have to rely on memory or scribbled notes to write a trip report back at the office.

An interesting use of hand-held recorders was instituted in a South American oil refinery, where a group of 13 nationals had to make daily inspections and submit reports in English at the end of the day. As is often the case with persons using English as a second language, they spoke it better than they wrote it.

After they had been exposed to the Murray System of organizing information, their manager bought each employee an inexpensive recorder and warned that secretaries would no longer be available to transcribe their handwritten notes. According to the manager, the improved quality of reports (including the use of English) was evident right away, and the time spent preparing reports was dramatically reduced.

A Cost-Cutting Measure

From an institutional standpoint, the most important reason for encouraging writers to dictate is probably financial. Dictation saves big dollars. From an on-the-job writer's point of view, generating these savings is an easy way to increase his value to his organization.

Some years ago a research organization with 240 professionals calculated that dictation alone would save over $1.1 million annually in productive research time gained. As a result, the company bought dictating machines and had a sales representative demonstrate their use to all participants in technical writing workshops.

If we brought overhead costs more in line with, say, 1985 realities, the savings would be even more impressive. Consider a research organization of 200 scientists and technicians who spend an average 20 percent of their time writing reports in longhand. If they work 240 days a year, the organization as a whole is dedicating 9,600 man-days annually to pencil-pushing.

Dictation can save about 75 percent of that time, or 7,200 man-days—equivalent to adding 30 more professionals to the payroll, with no salaries to pay, no fringe benefits, and no recruiting or training costs. At an overhead cost of $400 per day to keep each professional on the payroll, the value of productive time gained would amount to $2,880,000 annually.

If you are a confirmed pencil-pusher, try keeping records for a month. The accumulated time and calculated costs of your writing in longhand should convert you to dictation right away.

The Computer Connection and Mandatory Dictation

As this book is written, the time is practically at hand when professionals will no longer have an option; dictating will be a required skill.

We can already drop the pencil and bypass the secretary if a word processor is available. Some professionals who can type as fast as 40–50 words a minute prefer the word processor to the dictating machine. At the present time, this may be a valid reason for not dictating. It is not a particularly cost-effective practice, however; a professional makes a pretty high-priced typist.

But the most important computer development in this area is speech recognition, or voice recognition, whereby a computer recognizes, as input, words spoken by a particular human voice. Some computers can already respond to a limited number of spoken commands. American, Japanese, and European firms are racing to be first with smarter-listening computers, for the market potential is immense. Clearly, it won't be long before we have the capability of bypassing the keyboard entirely.

When that day comes, it is not hard to predict another revolution in technical and business communication and information-handling. The office of every person who has to write will be equipped with a microphone connected to a computer, and management will insist that everyone learn to dictate. Secretaries will be freed for other responsibilities. Typing will become a much less common way of entering data into a computer or word processor.

The prudent writer will do well to prepare now for these inevitable changes.

HOW TO BE A
GREAT DICTATOR

Now that your ideas are well-organized and approved (Steps 1–6), you can talk confidently about them. So talk: dictate. For short documents and letters, work from an indication of your answers to WHAT, WHY, HOW, and NOW WHAT. If you are writing a lengthy memo or report, start with the Summarizing Topic Sentences in your Expanded Outline. Explain, support, or develop each idea; then go to the next ready-made starter sentence.

Nobody can dictate successfully without first organizing ideas.

If you plan to use a dictating machine, here are some helpful tips.

(1) Take a few minutes to become familiar with the recording unit you are going to use. If it requires batteries, are they fresh? Are the tape size and speed compatible with those of your typist's transcription equipment?

(2) Keep the microphone a few inches from your mouth, as you would a telephone. Speak slowly and distinctly.

(3) At the beginining of the tape, instruct your typist (call typist by name if you know it).
Tell WHAT the document is (letter, memo, report, etc.). Final copy or double-spaced rough draft?
To WHOM is it addressed? (Spell out names as necessary, or give report title and number, memo heading information, etc.)
Give HOW information:
HOW many copies? (Spell out names of recipients.)
HOW should tables, figures, and equations be treated? (You may state that these are attached for insertion or instruct the typist to leave space for later insertion by author.)
HOW should the document be laid out? Block paragraphing or indent five spaces? (Such matters are often specified in a secretarial manual.) If it's a report, you might say something like, "Center first-order headings, caps and underlined. Bring second-order headings to left margin, upper and lower case, underlined."
If you have a deadline, tell the typist.

(4) Feel free to ask for a rough draft, and plan to edit it. Two drafts call for a little more typist's time, but they save professional time, which costs more. If the typist has a word processor, fresh drafts are easy.

(5) After your instructions, leave a blank space on the tape for belated changes or insertions.

(6) Now dictate the body of your text.

(7) If you're in doubt about what you've said, back up and listen. If you don't like what you hear, dictate over the offending passage, erasing it.

(8) If punctuation is important, call it out as you dictate.

(9) Spell out technical terms and proper nouns as you dictate. (If the typist is not acquainted with your terminology, you may wish to instruct that these terms be left blank for later insertion by hand. Paleontologists, for example, often do this.)

(10) If you play your message back and discover you've left something out, just dictate your correction at the end of the message ("Correction: after Paragraph 2, insert the following: . . ."). Then go back to the blank space you left at the beginning of the tape and warn the typist to go to the end and listen for corrections. (If your typist has a word processor, this little trick will not be needed.)

(11) Don't worry about halting and correcting while you're learning. And don't try to sound professional. You are professional.

EXCUSES, EXCUSES

Considering the compelling reasons for learning to dictate, it is surprising how many professionals still insist on writing out reports and correspondence in longhand. They offer various "reasons," all of which, though sincere, are questionable:

■ There are 15 people in my group and just one secretary, so only the boss gets to dictate.

Have you tried a dictating machine? Transcribing a tape takes less secretarial time than deciphering 15 different styles of handwriting.

■ Our organization is on an economy binge and won't buy dictating equipment.

Odds are you haven't asked. And if you have, did you marshall your statistics on savings and prepare a fully justified recommendation?

■ Our secretary doesn't like machines—refuses to transcribe tapes.

Who is in charge? Maybe you need a new secretary. Good secretaries are not only eager to learn new skills and willing to cooperate; they are also knowledgeable and patient instructors in what to do and what not to do in dictating.

■ I compose better in longhand, when I can study what I've said.

Does that mean you write it out in order to think it out? (Not a very logical method of organizing.) And have you considered that you can play back your message on tape and revise it by simply dictating over the part you want to change?

■ My supervisor doesn't believe in dictation—doesn't think it saves time.

Catch him in a meeting, give him a paragraph from a report, and time him as he reads it aloud. Then time him as he self-consciously writes it out in longhand. An engineer in one of our classes convinced his boss that way, using only a long report title.

■ I don't write enough to justify running around to find a machine, and what I do write is very short and simple.

True—if you write, say, only one or two short memos every couple of weeks. But considering the changes computer technology will inevitably introduce (as well as the fact that your job description may change some day), it wouldn't hurt to practice dictation in your spare time.

THE LAST STEP

Back off, cool down, then start editing (Step 8).

The most uncontrollable human urge

stems neither from sex nor greed.

It is the urge to ~~revise~~ ~~FIX~~ ~~improve~~ ~~Modify~~

another person's ~~writing.~~ ~~TEXT~~ ~~prose~~ ~~Documents~~ ~~memos~~

STET!

Step 8
Edit the Typed Draft

Letters and short memos usually do not require much editing following Step 1 (the analysis), Step 2 (organization of the essential message), and Step 7 (dictation). With practiced dictation, these brief documents should come back from the typist ready to be signed and sent off. On the other hand, few people can write a long memorandum, a formal report, or a 15-minute speech without editing a draft.

This final task should be relatively fast and painless; editing a well-organized document takes only a fraction of the time and effort spent on material that has to be spliced, patched, and rewritten. Moreover, it's hard to make a patched-up job look like anything but what it is.

Efficient editing, like efficient writing, requires logical procedures. We cannot effectively do everything at once; the job needs to be divided into manageable steps with specific objectives.

SIX TIPS FOR EFFICIENT EDITING*

(1) Cool Down Before Editing

It's often said that the draft should be allowed to cool, but it's the writer who needs the cooling period—to recover from the fever of creativity. A little distance from that hard-won prose does wonders for objectivity.

No doubt in your editing experience you've recognized a sparkling passage and thought, "Hey, that's really good!" You may also have had moments when you looked at the cold prose and wondered, "Now what did I mean by that?" Writers ought to be able to react in both ways—viewing their writing as a stranger's prose—before starting to edit.

(2) Edit First for Organization Only

You wouldn't wash the windows in a house that was about to collapse; nor should you repair sentences until a document is structurally sound.

---

*We cover editing for clear and correct English (the so-called "clear-writing techniques") in Part III.

Overall organization should not present problems by this time because it has been based on guidelines that force the **form** to fit the **people** involved. It should be possible now to scan Summarizing Topic Sentences of major units, skimming lightly over details, and get a coherent story.

If that is not possible, some major unit is probably backward, lacking a good Summarizing Topic Sentence. Stop where the story seems to get lost—where you have to dig into details to get the point or the connection—and find the missing idea. It may be at or near the end of the unit, or it may not show up anywhere.

In either case, when good Summarizing Topic Sentences are inserted at the beginning, several sentences or whole paragraphs can usually be scratched, reducing length appreciably without harming the message at all. This agreeable possibility is demonstrated in the following examples, where the Summarizing Topic Sentences obviously make the trite introductory information superfluous:

Introductory Topic Sentence	**Summarizing Topic Sentence**
(a) As you know, our decontamination work has been hampered by the condition of our decontamination cabinets.	We recommend that three Brand Z safety cabinets be purchased at a cost of $4,500 each.
(b) A survey was recently conducted to determine the effect of company innovations on drilling costs last year.	According to our recent survey, company innovations reduced drilling costs 11 percent last year.
(c) At our last safety meeting, Ms. Overvue stressed our continuing interest in reducing accident rates in the organization.	At our last safety meeting, Ms. Overvue reported that lost-time accident rates increased in all branches last year, as follows:
(d) A study was initiated with the objective of developing a standard procedure for raising the surface temperature of producing-well fluids.	This report presents a new standard procedure for raising the surface temperature of producing-well fluids.

A comparison of the sentence **skeletons** (underlined) in the two columns above makes clear that the discussions to follow will be entirely different, in accordance with the theme established by the skeleton. Moreover, the skeleton can clearly be informative, as in the right-hand column, or of little interest to anyone but the author, as in the left-hand column.

(3) Prepare for Sentence Editing

While scanning for defects in organization, flag awkward or unclear sentences with question marks in the margin. But don't stop to revise sentences. It may be hard to pass them by, but you can do a better job of repairing them once you have an overview of the whole story.

(4) Edit Sentences and Polish

Go back to the sentences you flagged as you edited for organization. Now is the time to look for the best choice of words; correct the grammar and usage; break up long sentences; combine short, choppy sentences; provide missing transitions; and check spelling and punctuation. You will find techniques for clear writing at the sentence level in Part III, along with a review of grammar and usage problems.

(5) Take Advantage of Computerized Help

Computers or word processors can take much of the pain out of correcting a final draft if they are equipped with a ''spelling checker'' program. A good spelling checker can do the following:

- Locate spelling errors and unrecognized combinations of letters
- Hyphenate correctly (especially helpful if you want justified right margins)
- Flag misplaced, excess, or ''orphaned'' punctuation
- Spot unbalanced quotation marks, parentheses, and brackets

An unrecognized combination of letters may, of course, be a valid word or acronym that is not in the spelling checker's dictionary. In that case the program should give you the option of adding it to the dictionary.

An error that a spelling checker cannot catch is a properly spelled word that is nevertheless a typographical error. For example, if the typist types "A rose by any other dame would small as sweat," the program will not detect any of the three mistakes. (For that kind of help we may have to wait for the next generation of intelligent word processors.)

(6) Plan Layout for Readability

This final task is discussed below.

TIPS FOR IMPROVING GENERAL READABILITY

The term *readability* means more than the original "reading ease" as measured by popular readability formulas. Most of these are based largely or entirely on sentence length and frequency of polysyllabic words. As important as length of words and sentences is, short sentences and one-syllable words do not guarantee readable writing. Research has demonstrated that comprehension and retention are controlled to a significant extent by the physical appearance of a document as well.

In other words, the layout of a report can make it dull, dense-looking, and hard on the eyes; or it can render a discussion with high technical density visually inviting, easy to absorb, and hard to forget.

(1) Plan for White Space

White space is a journalistic term for the blank areas on a page. To appreciate its value, compare a page of want ads with a professional advertising layout. Tiny type and crowded, wall-to-wall text make want ads uninviting and slow going. The term *fine print* is almost a code term for "the part of a document that nobody reads"—and with good reason. You have to have a special motive for tackling fine print. By contrast, the kind of advertising that sells invites attention by its unconfined prominence within white space. Similarly, on their news pages magazines use ample margins, space between paragraphs, and boldface headlines (to say nothing of color) to entice you to read. Some on-the-job writers reject the notion of salesmanship in connection with scientific or technical prose, believing that good work sells itself. But the truth is that if we can't sell a new idea or new technology in print, it may never get the hearing it deserves. This has been demonstrated over and over again in industry's research organizations.

If you skipped through or over the last paragraph, please go back and read it. We crowded it just to show how insufficient white space turns readers off.

Here are three techniques that increase white space—without loss of dignity or professionalism:

Avoid narrow margins

In your instructions to the typist, specify margin width. Anything less than one inch (2.5 cm) on the right, left (after binding), and bottom crowds a typewritten page. Ideally, at least 1¼ inches (3 cm) should be allowed. To be esthetically pleasing, very short letters (only a few lines) should have even wider margins. They can also be double or one-and-a-half spaced.

One situation that does justify a bit of crowding is an opportunity to confine a letter or memo to one page. There is something highly attractive to busy readers about a communication with the author's signature at the bottom of a single page. The limited length is so obvious that they are likely to read it immediately. (Besides that, carrying a line or two to a second page wastes stationery.)

You might also decide on narrower margins for a very long report that would discourage readers by its formidable bulk. Reducing margins from 1¼″ to ¾″ on 8½ × 11-inch paper can cut the number of pages by nearly a third. But before you take this step, look for ways to reduce an overweight report by such strategies as printing on both sides of the page and consigning backup data to detached supplements.

Paragraph frequently

Nothing is more discouraging to a reader than heavy blocks of print that go on and on without the relief of white space. Some writers believe that once a subject is introduced, it must be exhausted before it can be interrupted. A suggestion that paragraphing would be helpful brings forth the unanswerable question, ''What's the rule?''

There are no rules. A paragraph can be ten sentences long, or it can be one short sentence, if that's all there is to say about the topic. It's a matter of judgment. The best way to judge is to consider what would help the reader separate subjects, and within subjects, how to help him separate topics.

If you want to stress a thought, that's another opportunity to paragraph. The president of a company, in reply to an employee contributing to an "earn-from-your-ideas" program, started with this one-sentence paragraph:

You have an excellent idea.

Another writer ended his report on an inconclusive study with this one-sentence paragraph:

It's a tough problem.

In paragraphing, as in anything else, we can have too much of a good thing. A whole string of one-sentence paragraphs would read like machine-gun fire, rat-a-tat-tat; relationships between ideas would be shot to pieces.

List within text

Another good way to create white space is to list instructions, how-to information, or inventories of items. Compare these two arrangements of the same information:

To ensure a reliable product, you should start with raw materials from Plant 12, blend them for at least 40 minutes, keep the temperature above 160°F during cooking, make sure that you have used sufficient Type M catalyst, and allow at least 18 hours for cooling.

To ensure a reliable product, you should:

- Start with raw materials from Plant 12
- Blend them for at least 40 minutes
- Keep the temperature above 160°F during cooking
- Make sure you have used sufficient Type M catalyst
- Allow at least 18 hours for cooling

Being easier to read, the version on the right is easier to grasp and easier to remember.

(2) Use Informative Side Headings

Headings placed at or in the left-hand margin have three important functions:

- To orient the reader by signalling the end of one topic and giving some idea of what's coming next
- To provide handy signposts for readers seeking a particular topic under a general subject heading
- To introduce white space for readability

Very simple **label headings,** like "Readability" or "Results" or "Discussion," are better than nothing, but they are rarely specific enough, except in the minor subdivisions of a section. Descriptive modifiers such as adjectives, nouns, and phrases help, but you have to watch out for trains of modifiers.

For example, this section falls under the side heading "Tips for Improving General Readability," in which "Tips" is clearly what we're talking about. Most technical writers would change that to "General Readability Improvement Tips." With this long train of modifiers preceding "tips," the reader wouldn't know what we're talking about until he worked his way from the "locomotive" to the "caboose":

General *readability?* No.
General readability *improvement?* No.
General readability improvement *tips?* Ah, there it is!

That sort of string, or train, of modifiers is actually a mini-version of suspenseful writing. We see many ludicrous trains of modifiers in nearly every technical journal published. Try these from the world-class collection of Professor Robert L. Bates of Ohio State University:

Liquid Metal Fast Breeder Reactor Spent Fuel Shipping Cask

Total Loss Control Management Conference

When the train is composed of nouns only, it's even harder to tell what the writer is going to talk about. (See p. 151 for a discussion of Noun Trains.)

Informative side headings are those containing a verb or verb form. While they are not always suitable, they can be used effectively in many texts. Notice how much more they tell you than labels do:

Label Heading	Informative Side Heading
Readability	Techniques for Improving Readability
Starting Salaries	Starting Salaries Rise Sharply
Field Trials (or Plans)	Field Trials to Start in March
Corrosion Inhibition	New Pipes Resist Corrosion
Test Results	Tests Confirm Chip's Reliability
Recent Progress	Playback Model Nearing Completion

Although the informative headings are longer than the labels, they take no more vertical space. Brevity is indeed a virtue; but when we weigh information against length, the choice here is clear.

You will also note another advantage of the informative side heading: it can feature NEWS items or main points where they can't be overlooked. If a reader does nothing more than scan the table of contents or flip idly through the pages of a report, informative headings give him the essential parts of the message. And that may just spark enough interest to entice him into reading parts of the text he would otherwise avoid.

(3) Use Comparisons and Examples

These techniques can make your ideas come alive in the reader's mind. They can clarify complex or obscure concepts. Comparisons can be as brief as a simile (the Model 6 is *like* the old Model 3A, with an improved keyboard) or metaphor (the project turned into a witch hunt). Examples can be as detailed as some of those in this book (see Step 5, for instance).

(4) Preserve Your Individual Style

Personal style is the sound of a human voice in prose; it comes through to the reader in both choice of words and structure of sentences. You made it yours by dictating. Don't lose it when you edit by trying to make fresh, spontaneous expressions more "impressive."

There are really only two things you need do to develop a readable style:

- Dictate your rough drafts (see p. 113).

- Edit for brevity and clarity so that no one will be tempted to tamper with your writing.

If you are successful in these two things, you won't have to worry about style. Your readers will be grateful for the conciseness of your writing. And if they think about it at all, they will probably admire the ease and "transparency" of your style.

(5) Run a Check with Readability Formulas

The writer side of us is often paradoxical: intensely proud and independent, yet deeply insecure and in need of support. For that vulnerable side, when it shows up, readability formulas offer unthreatening aid. They are of course somewhat limited, for what they tell us is chiefly whether sentences are too long on the average or are laden with too many "big" words—or both. But at times that is all we need to know in order to improve the written product.

A widely known readability formula is that of Rudolf Flesch, published in 1949. In 1978 General Motors Service Research programmed the "Flesch Reading Ease Index" for the GE Mark III time-share computer. The program is relatively short and not difficult to modify for other computer systems. The retail software market already offers other programs to help us shorten sentences and replace polysyllabic jawbreakers.

Another widely used formula is Robert Gunning's **Fog Index.** Like Flesch's formula, the Fog Index is based on both sentence length and polysyllable frequency. It contains a factor that converts the numerical score (index) into the number of years of education the reader would presumably need to read the sample with understanding—for example, writing with an index of 16 could be handled by a college graduate. Anything with an index over 12 is considered "foggy" writing, but we have seen scholarly reports with indices as high as 24—virtually unreadable.

Readability formulas can, of course, be misused. They are di-

agnostic tools, not clear-writing techniques. That is to say, trying
to reach an easy-readability rating set as an arbitrary goal can lead
to an oversimplification that actually blocks the flow of ideas. Former Secretary of Education Terrell Bell attributed what he called
the "dumbing down" of educational material—and the attendant
decline in S.A.T. scores—to such a use of readability formulas by
textbook publishers, at the insistence of educators.

Scientific and technical writing can also be "dumbed down"
by unthinking conformity to the little-word, short-sentence ideal,
but one need not get caught in this trap. (See Part III, p. 141, for
the kinds of big words that create unnecessary—often preten-
tious—complexity and the kinds of long sentences that deserve
pruning.)

TIPS FOR THE
SUPERVISORY EDITOR

Editing the work of a touchy writer can be a thankless chore,
and all writers—including even thee and me—can be touchy about
their words. We will defend them to the last ditch (or the last
passive verb).

Professional editors have an advantage over supervisors in their
specialized education and experience, not to mention a job description that makes them "experts." A supervisor-turned-editor, with
no claim to language expertise, has a hard row to hoe indeed. He
may be highly esteemed as a checker of fact or policy, but it does
not follow that his attempts to correct grammar and syntax evoke
the same respect.

One should not assume that all supervisors are poor editors
because they are not professional editors. Many people have a
good ear for language and can detect and correct discordant sounds
without knowing why. Morever, editing is in part an exercise in
people-handling, which means that experienced supervisors have a
good background for dealing with sensitive writers. Experienced
people-handlers will instinctively phrase their criticism something
like this: "You've got a really good section here; I'd like to see
you move it up front for the impact it deserves," rather than,
"You've certainly managed to lose me here—what's this all
about?"

There is a Golden Rule of Editing that most supervisors would
have difficulty following: **never change a writer's words unless
you can explain why.** However, some writers do such a miserable

job that they must have help, even if the supervisor with no claim to expertise is the only help available. Furthermore, some writing simply could not be released to represent an organization's standards of literacy. Qualified or not, willing or not, a responsible supervisor has to try.

The question, then, is how to ease that task and reduce the valuable time diverted from technical and executive jobs to the editor's job. We believe the techniques and procedures offered in this book will be as helpful to supervisor-editors as to professional editors. In addition, we offer the following special suggestions to supervisors.

(1) Try to Ensure that Incoming Drafts Are Well-Organized

Encourage writers to follow Steps 1 through 8 as required by the type and length of the writing job. This will cut both the amount of editing necessary and the time consumed, and it will produce a better document than editing alone ever could.

The most time-consuming mistake any editor can make is to start tinkering with sentences when the real problem is botched organization; every change seems to lead to another, and pretty soon more harm than good has been done by editing. It really is a waste of time to get bogged down in cosmetic changes when re-organization will render them irrelevant.

Once a report is properly organized, the Essential Message will be highlighted throughout, and gaps in logic, discrepancies, and apparent contradictions that need fixing will be less likely to occur. Moreover, if the text has been dictated (or talked into a keyboard), the language will be simpler and more natural; ideas will be less densely packed; and meanings, even if couched in ungrammatical terms, will be easier to understand and thus easier to repair.

(2) Support Pre-writing Conferences

We urge you, in your own best interest, to assure your writers that you will be available for Pre-writing Conferences (Step 6). Let them know you are willing to meet with them *before* they spend days or weeks composing a text that you may have to disapprove or modify significantly.

"My boss would never agree to that," many writers tell us. "He's always too busy." But in fact we have never, in all our seminars and interviews, found a supervisor who would refuse.

They consider the time saved, the improved morale, and, most of all, the improved products that derive from the Pre-writing Conference to be well worth the effort of participation. It clearly beats having to wade through a disorganized report, struggle to pin down what makes it so bad, develop constructive help with the problems, and then consult at length with an unhappy writer.

(3) Expect Decent Editing from Writers

In organizations where a great deal of supervisory editing is done routinely, writers tend to develop a **care-less** indifference toward their writing jobs: "Why bother? It's all going to be changed anyhow!" And they turn in work that no supervisor should have to accept. Sometimes it seems that the more editing you do, the more you have to do.

If the situation is very bad, we recommend training. But if the problem is just sloppy grammar, a writer can help himself. Numerous good books on grammar and usage are available to refresh the careless writer's memory (see Selected References). We also offer basic help in Part III of this manual.

(4) Acquire References that Editors Rely on

Nobody carries all the answers in his head; knowing where to find them is the mark of an expert. We consider the following a minimum:

The Elements of Style (Strunk and White, 1979). A classic condensation of the principles of good writing in English. Belongs on every editor's desk.

A Dictionary of Contemporary American Usage (Evans and Evans, 1957). An authoritative guide to grammar and preferred usage, in dictionary format. This book, now in its 14th printing, is witty, entertaining, and easy to use. We consider it a basic reference.

The Careful Writer: A Modern Guide to English Usage (Bernstein, 1977). An excellent reference on grammar and usage in dictionary format. Bernstein is not only an entertaining writer; he is also a help on subjects that many style and usage guides omit, such as the use of prepositions (particularly good for writers using English as a second language).

The American Heritage Dictionary of the English Language

(1979). This is one of the few dictionaries to include "Usage Notes" indicating *preferred* spellings, pronunciations, and senses of use. An abridged version is available in paperback, but the unabridged dictionary is more useful for editorial work.

Roget's International Thesaurus (1977). This complete book of synonyms and antonyms in American and British usage is a must for both writers and editors.

A Manual of Style (University of Chicago Press). Guidance in usage of capitalization, punctuation, italics, notes and footnotes, abbreviations, and other matters of format.

These references are indispensable to supervisors like a participant in one of our seminars who said, "Frankly, grammar has never been one of my strong points. I try, but my people keep telling me I've changed their meaning."

Another supervisor in the group explained how he handled that problem: "I just put question marks and cryptic phrases in the margins, and ship the whole thing back to the author with an encouraging comment—something like, 'Good report, Joe, but it needs more work.'" That approach is no more constructive than the unfortunate "improvements" of the first speaker.

Still another supervisor, an engineer with no claim to expertise in language, had a better solution. She studied some authorities, listed the errors she encountered most frequently in editing, and made a point of explaining her changes to writers. "It takes time," she agreed, "but I consider it a training aid." Moreover, she said that having the answers greatly enhanced her authority as an editor. In time, it will also reduce the burden of editing. Remember the old Chinese proverb:

Give a man a fish
and you feed him once;
Teach him to fish
And he can feed himself for the rest of his life.*

A final note on authorities: They often disagree. They argue among themselves. They understandably base some of their advice on their own preferences. This means that if you shop around, you may find an authority who supports your preference.

---

*For further thoughts on this matter, see the section on ideas for institutional editors (p. 137).

(5) Practice Restraint in Matters of Personal Style

First, what is personal style (as contrasted with manuscript, typographic, or usage style)? Here are three definitions, proceeding from the lexicographic to a Strunkian aphorism:

■ The way in which something is said or done, as distinguished from its substance; the combination of distinctive features of literary or artistic expression, execution, or performance characterizing a particular person, people, school or era. *(The American Heritage Dictionary*)*

■ Style in writing and speech is a blend of one's personality, temperament, and individual peculiarities. It is a characteristic of good writing that delivers readers from monotony, dullness, and boredom. *(Geowriting: A Guide to Writing, Editing and Printing in Earth Science,* 1984)

■ . . . style *is* the writer, and therefore what a man is, rather than what he knows, will at last determine his style. *(The Elements of Style,* Strunk and White, 1979)

No wonder people are at their touchiest when they think you're tampering with their personal style! They have a right to be. Editors have an ethical obligation to avoid imposing their own style on another's writing. Few would want to, anyhow, for a distinctive style is always refreshing to come upon during the work-day's reading. But transgression is all too easy when a sentence is so poorly written that it has to be reconstructed entirely.

There is one occasion, though, when you can ethically impose your style on another: if a subordinate is writing for your signature, you have a right to make the words sound like yours. It would be unrealistic, of course, to insist that subordinates get the sound of your voice into *their* prose—unless you are fortunate enough to have a ghostwriter on your staff.

The old saying "If it ain't broke, don't fix it" sometimes applies to supervisory editing. When you do make seemingly arbitrary changes, replacing a subordinate's good word with your "better" word, explaining the reason for it is a particularly constructive gesture.

---

*See Selected References (p. 255).

**IDEAS FOR
INSTITUTIONAL EDITORS***

What follows is proposed as a workable solution to two problems: (1) how to reduce the length of reports and articles without sacrificing desirable content, and (2) how to get the author to cull out and organize his own raw material effectively before he starts to write. Neither editor nor reader should have to slog through prose thickets armed with a yellow Hi-Liter® pen in a time-consuming search for gems of wisdom.

This labor-saving approach for editors is not really like pulling a Tom Sawyer. (You'll recall that Tom conned his buddies into whitewashing the fence for him by convincing them the job was a privilege and a delight.) Rather, editors—like dentists and physicians—should strive to work themselves out of part of their job with some preventive medicine.

This philosophy goes far beyond handing each author a sheet of proofreader's symbols, because for many documents that would be comparable to offering Band-Aids® for a broken leg. The analogy is more precise than it sounds: the problem with a great deal of scientific writing is structural. When the organization is faulty, no amount of superficial treatment is going to correct it.

This problem invariably shows up as buried NEWS. Resurrecting it at the editing stage and reorganizing the supporting material to follow the NEWS can be an involved task. It wastes time, duplicates effort, costs money, and strains relationships. The editor is stuck with time-consuming tasks the writer should have handled himself:

- Analyzing the project's background (original objectives, political or corporate restrictions, changes in either project objectives or reporting objectives, etc.)
- Determining the author's objectives in writing
- Identifying the various readers' needs and interests
- Identifying pre-reporting problems that were swept under the rug

Then follow the formidable tasks of re-outlining, reassembling, rewriting, and finally, of convincing a defensive author that "It's

---

*Hugh Hay-Roe, "Care and Feeding of Authors," Adapted from the newsletter of the Association of Earth Science Editors, *Blueline*, v. 16, no. 1 (1983) pp. 4–6.

really an outstanding report, but . . . ." Such editing is not first-aid; it's major surgery, and given the reaction of many authors, it suggests that malpractice insurance for editors might not be a bad idea.

Everyone can benefit from more efficient procedures that will boost productivity. The ideal solution is to invest some editorial time in training writers, and the best way to start on this is to schedule the first editor-author conference *before* the author gets involved in the emotional business of authorship.

Give the author on-the-job training in organizing his ideas; go over Steps 1 and 2 with him. Elicit all the information needed for logical organization. Then lead the writer through Steps 3–5 to generate an expanded outline supported by relevant illustrations.

The first conference with the author can take as much as three hours, but the next one usually takes no more than half that time. After such conferences on a couple of reports, the author, won over, comes into subsequent meetings with expanded outlines ready for the editor's suggestions or revisions. Reorganizing an expanded outline is easy, and the author is grateful that you haven't—in his mind—desecrated his handiwork. Further, this editorial input can ensure a shorter, more harmonious Pre-writing Conference (Step 6) with supervisors and managers.

We don't propose training authors in order to encourage indolence on the part of overworked editors; nor will it jeopardize the editor's job security. It just makes sense for the editor to spend more time upgrading writers and less time upgrading their writing. In the long run, this approach will increase both the value and the prestige of the editorial position.

Part III

Editing for Clear and Correct English

EVOLUTION OF A MASH NOTE

We can imagine, at the dawn of human pre-history, a caveman whom we'll call UGG, wanting to communicate an important idea to his girlfriend, Ogga. At first he could manage only a very telegraphic message, not terribly grammatical:

In sentence diagramming, we show this above a baseline called the skeleton line, with verb separated from subject and object:

UGG	LUV	OGGA

Later, as his spelling and linguistic skills improved, Ugg decided to brag a little — with an adjective. He called himself "great Ugg":

UGG	LOVE	OGGA

GREAT

But this got him nowhere with the lady, so (while learning the singular form of the verb) he attempted flattery with another adjective. He called her "beautiful Ogga":

UGG	LOVES	OGGA

GREAT · BEAUTIFUL

Still later, moved to heights of eloquence, the budding writer included adverbs to modify *loves* and *beautiful*:

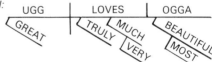

UGG	LOVES	OGGA

GREAT · TRULY MUCH VERY · BEAUTIFUL MOST

We now have a diagram of the complete sentence, "Great Ugg truly loves most beautiful Ogga very much," with its six modifiers below the skeleton line. But no matter how complex the sentence becomes, we are still "programmed" to anticipate its essence in the skeleton. If we don't find it there, we may be led astray.

Part III
Editing for Clear and Correct English

> I beseech you, graduates: Join in the guerrilla war against pollution of our public language. Strike in the active voice. Aim straight for the enemy: imprecision, ambiguity, and those high words that bear semblance of worth, not substance.
>
> —Bill Moyers*

By this time you have a basically sound product, well worth polishing (if necessary).

Since your document in expanded outline form has been approved up the line, there is little chance that editing will be wasted effort; you can work with confidence. Nor do you face any major cutting, patching, or rewriting; the whole has been properly organized. And since the document was dictated or "talked into the keyboard," it has the sound of a human voice (yours, happily) in the prose. Thus sentences will be less contrived; verbs will be more vigorous; language will be plainer; and tone will be easy, natural, and familiar to your ear.

Finally, with distance from the labor of creation, you will see your writing as a stranger's prose. Altogether, this means that the sentence problems that have crept in will be easier to spot, easier to understand, and easier to fix.

Most writers make only a few of the common errors in writing—certain grammatical errors; particular misspelled words; questionable punctuation; habitual use of weak or distorted verbs; wordiness; addiction to big words and fancy language; unclear pattern (lack of transition signals); overly long and loaded sentences or too many short, choppy ones. Knowledgeable editing can identify these flaws, but understanding will help to wipe them out of one's habitual way of saying things. This ability to edit with understanding is the purpose of the following sections.

---

*Address to the graduating class, Lyndon B. Johnson School of Public Affairs, University of Texas at Austin, May 1985; excerpt from the Houston *Post,* May 25, 1985, p. 3B.

141

Apply Clear-Writing Techniques

Making your message clear is not so difficult if you have the means. How many ways do you know to clarify your writing? Here are 18 reliable techniques.

PREFER THE ACTIVE VERB

Verbs are the life and vigor of your writing. They can project the image of a forceful worker or an uncertain plodder. The simple technique of using active verbs will do much for your professional image. Take advantage of the richness, strength, character, directness, and brevity this easy technique imparts to your prose.

Note that we say (in the heading above) *prefer* the active verb; passive verbs are also useful and necessary. We need them to emphasize the right subject, to make a transition from one topic to another, and to provide pleasing variety. The problem is that most writers use the passive form almost exclusively. Unfortunately, we have been taught since the days of devotion to Victorian modesty and humility that active verbs are not "objective." Accordingly, most of what we read is cast in that presumably objective passive form. We pick up the habit without recognizing it.

Voice is the most important characteristic of verbs in connection with clear writing. Verbs are said to be in the active voice **when the subject performs the action of the verb upon a direct object:**

> S   V   D O
> Figure 73 illustrates the capabilities of the new unit.

If the action of the verb is performed upon the subject of the sentence, the verb is said to be in the *passive* voice, and there is no direct object.

> S   V
> The capabilities of the unit are illustrated by Fig. 73.

That's the grammatical way of explaining voice. Perhaps an easier way is to look at the flow of the action. From the time we started putting nouns and verbs together as children, we have been programmed to follow that flow of action in the normal word order

The Power and Subtlety of the Verb

The verb is a strong yet delicate instrument for providing exactly the right nuance. With the proper verb, you don't need to rely on adverbs and adjectives. As an example, here is a spectrum of verbs to choose from when you have to ask someone (in writing) to do something:

[would] appreciate
[would] suggest, propose
prefer, would like
request, ask
recommend, urge
remind
advise
require
insist
demand

In using forceful verbs, don't overdo it; the results can be ludicrous. One author, striving to be dramatic, spoke of ". . . the pipeline that *slices* across the Syrian desert . . . ." Pipelines do not slice; being round, they have no cutting edges. They just lie there.

for our language. Normal word order in English runs from actor through action to receiver of the action.

When the flow of action is backward (passive), we perform two operations simultaneously: the eye reads from initial capital letter to final period, forward; but the mind, seeking "who did what," has to work in the reverse direction.

Let's consider the flow of action in this familiar quotation from Aesop's "The Fox and the Lion":

S   V   D O
Familiarity breeds contempt

actor ⟶ action → receiver

This is an excellent sentence. It takes only three words to express a big idea. The flow of action is straightforward. The verb is strong. And the whole idea is displayed in the skeleton.

Now let's turn it around to make it passive.

Contempt   is bred by familiarity.

receiver←—action←————actor

Reversing the flow of action has almost doubled the number of words, weakened the verb, and lost part of the idea (the actor) from the skeleton line. Compare "familiarity breeds contempt" with "contempt is bred."

Strong Sayings, Passive-fied

Here are examples of what passive verbs can do to strong statements:

Liberty should be given to me or death should be given to me.

The heart is made to grow fonder by absence.

The coming of the British has been observed! (Paul Revere)

These truths are held to be self-evident by us.

Let cake be eaten by them.

In scientific and technical writing, however, we do worse: We tend to turn passive verbs into another form that damages expression—the "distorted verb."

CHECK FOR DISTORTED VERBS

Returning now to the weakened version of Aesop's saying, let's turn that passive verb, is bred, into a noun, breeding, and make it the subject of the sentence. We will then have robbed Aesop's sentence of a verb entirely:

Breeding of contempt _____?_____ by familiarity.

What can we do for a verb? What's your favorite? How about is accomplished? (Try reading a few passages from any technical report and count the number of times this verb is used, along with its mates, was carried out and is attained.)

S   V
Breeding of contempt is accomplished by familiarity.

receiver ←——————— action ←———————actor

Now the flow of action is backward; the original three words have become seven; and nearly all of the idea is missing from the skeleton—buried in modifiers. Compare "Breeding is accomplished" with "Familiarity breeds contempt." One can do worse than that, though, without trying very hard. We have another word cherished by technical writers: accomplishment. It is all too easy to turn the passive verb is accomplished into the noun accomplishment:

Accomplishment of the breeding of contempt ____?____ by familiarity.

What verb now? Writing in a rush, we've lost track of the sentence structure. But we do sense that we must have a verb, so we fill in hurriedly with another old standby: is achieved.

Accomplishment of the breeding of contempt is achieved by familiarity.

Now all sense has been lost from the skeleton of the sentence. Compare "Familiarity breeds contempt" with "Accomplishment is achieved." We have nonsense. We also have 11 words in place of 3.

What sensible writer, you might well ask, would make such a mishmash of a simple idea? The answer is that nobody would do it deliberately, but everyone is likely to do it unconsciously. We come across that construction so often in reading that we become inured to the pompous sound of it. Here, from a wealth of examples mined from technical reports, is a small sampling:

Passive and Distorted	**Active**
Displacement of serum by air occurs. (6 words)	Air displaces serum. (3 words)
Improvement of the techniques was accomplished for the **provision** of **reductions** in cost	We modified the model and simplified procedures to reduce costs. (10 words)

Passive and Distorted	Active
through **modification** of the model and **simplification** of the procedures. (23 words; additional distorted verbs boldface)	
Detection of leaks in the line can be accomplished with this device. (12 words)	This device detects line leaks. (5 words)
Improvement of the drillstem-testing equipment was carried out by this group. (12 words)	We improved the drillstem-testing equipment. (6 words)

Verbs can also be distorted by turning them into adjectives:

Verb Turned into Modifier	Good Verb
Flowing water moves through the passages. (Of course.)	Water **flows** through the passages.
Varying solubilities were noted (or were measured, or existed).	Solubilities **vary**.
Throughout the operation, **increasing** temperature was observed.	Throughout the operation the temperature **increased**.

Distorted verbs have a way of attracting unto themselves all kinds of pompous padding—another thing to watch for in editing.

CUT THE PADDING ATTRACTED BY DISTORTION

Going back to Aesop's sentence, let's see how much padding it might attract in the fashion of technical writing:

Accomplishment of the breeding of contempt is (currently in the process of being) achieved by familiarity. (16 words instead of 3)

Does that example look artificial? Then how about this:

> An evaluation of the dollar profits realized from research developments is currently in the process of being carried out by the entire Research Division.

Meaning, the author explained, that:

> The Research Division is calculating dollar profits from research developments.

Distorted Verbs—The Danger Signals

If you find yourself casting about for any of the verbs listed below, chances are good your best verb is already present, hidden in a word ending in *-ment, -ing,* or *-tion,* and used as the subject. Here are some weak verbs which are typical warning signs:

has been made	occurs
is carried out	exists
will be accomplished	takes place
was completed	was achieved
has been performed	was noted
was observed	is detected
has been done	was indicated
is seen	is involved

Those are the signs of possible distortion. Look out for them.

USE LITTLE WORDS
FOR LITTLE JOBS

Even if there are few distorted verbs to invite padding in your text, it may be padded with phrases built up of little words. Many such phrases can be reduced to a single word:

Little-Word Padding Phrases	Simple Meaning
to be in a position to	to
in order to	
with a view to	

Little-Word Padding Phrases	Simple Meaning
with a view toward for the purpose of	to
along the lines of in the nature of	like
in reference to with reference to with regard to with respect to	about
in the event that if it should turn out that	if
due to the fact that owing to the fact that in view of the fact that	since, because
will be of assistance in will be an aid to	will help
makes the statement offers the comment makes the observation	says, states, comments, observes

ELIMINATE
FALSE OBJECTIVITY

Many a robust verb has lost its vigor because of an author's attitude toward personal pronouns. The Phantom Scientist—parading as an objective, unbiased "it" but really a person with pen or pencil—pleads modesty and objectivity as he defends rules excluding first-person pronouns. The Timid Scientist always goes along. The Practical Scientist obeys these rules when he must.

The Phantom Scientist is responsible for prose like this excerpt from a journal prohibiting personal pronouns:

Claims for missing issues will not be allowed if loss was due to failure of notice of change of address to be received before the date specified in the above paragraph (July 19__).

Stripped of its third-person "objectivity," this weighty communication looks something like this:

If you have changed your address, please notify us by July 19__. Otherwise we cannot grant claims for missing issues.

Pronouns neither create nor destroy objectivity. If you think objectively, you will write objectively. If you are biased, be on guard, for your writing will reveal it. For instance, one rather stuffy engineer, trying to avoid the conceit of personal pronouns, wrote this:

It is extremely regrettable that acceptance of this challenging offer is prevented by prior commitment to projects for which adequate replacement would be impossible to secure.

Queried by an editor, he phrased a more humble regret (and a more honest one, he admitted) by using five personal pronouns:

I am sorry that I cannot accept this challenging offer, but I do not feel that my experience qualifies me for the job.

In addition to creating false objectivity, exclusion of personal pronouns may create ambiguity. In the following sentence, who did what?

Since the calculation was considered questionable, an investigation was initiated.

We or they considered? We or they investigated? To be clear in context, the sentence should have been written like this:

Since they questioned our calculation, we have started an investigation.

Caution: The personal pronoun is a useful one of the eight parts of speech, but it can be misused. In the following sentences, the use of personal pronouns would justifiably be criticized; they are superfluous. The verbs can be active without pronouns.

Active but Personal	**Active, Impersonal**
<u>We</u> have successfully developed measures for preventing corrosion where it was formerly impossible.	The Engineering Dept. has developed a reliable corrosion inhibitor.
<u>We</u> can minimize corrosion by using special metals.	Special metals minimize corrosion.
In this report, <u>we</u> describe X-ray data for . . .	This report describes X-ray data for . . .

Personal pronouns also invite criticism when they are overused. We do well to avoid them in describing what we did (procedures, for instance), where every sentence would begin with a *we*.

CUT UNNECESSARY TECHNICAL JARGON

The term **technical jargon** (the special language of a profession) is not necessarily pejorative. One short phrase may take the place of a paragraph. Scientists could not think, talk, or write without the special language of their discipline. In writing to peers, they should of course take advantage of the standard shortcuts offered by their specialized terms.

Unnecessary jargon, however, is a different matter. If it is used when plainer, nontechnical words have equal acceptance, jargon is offensive. And when used to impress nontechnical readers, it becomes rude gibberish.

The safe approach is to temper your technical language with common sense, recalling the readers analyzed in Step 1. If they are not all specialists in your field—

This Would Be More Appropriate	**Than This:**
Shear pins attach the blades to the sleeve.	The blades are releasably attached to the sleeve by frangible means.

Table I summarizes the results of soluble organic matter (SOM) and total organic matter (TOM) analyses.

Analytical results of SOM and TOM are summarized in Table I.

We sampled the nearshore sediments at 5-foot intervals.

The littoral clastics were sampled under conditions that would assure adequate coverage of the interval of interest.

Comparative tests reveal that the contents of the volatile memory cannot actually be transmitted at 2,400 bits per second.

Benchmark results reveal that a dump of the RAM buffer does not actually occur at 2,400 baud.

UNCOUPLE NOUN TRAINS

A noun train is a string of nouns used as modifiers for another noun at the end of the series. Bates*, who provides the first example below, calls the use of lengthy noun trains "nouniness" or "nounspeak." As we mentioned on p. 129, whether the modifiers are nouns or adjectives, in a long train they represent a mini-version of suspenseful writing. To help the reader, they should be broken up:

Before	**After**
Flue Gas Desulfurization Sludge Disposal Practices	How to Dispose of Sludge from Desulfurization of Flue Gas
Carbon-13 Cross Polarization Magic Angle Spinning Nuclear Magnetic Resonance Analysis	(Would you care to try this one?)

GET RID OF FLOWERY LANGUAGE

In some respects we are still struggling to fight free of Victorian fashions in flowery language. Probably few business people

*Robert L. Bates, "Nouniness and the Missing Hyphen," *Journal of Sedimentary Petrography,* v. 50, no. 3, (September 1980) 1022–1024.

today are capable of writing, ''In urgent acknowledgment of yours of the 25th inst.'' But there are writers around who can, without embarrassment, make statements like, ''Hence we deem it essential to foster effective government support for the maturation of comprehensive, goal-oriented strategies that will enable fundamental scientific research to advance on multiple broad fronts as the ensuing century is approached . . . .''

Most of us need not be admonished to avoid that kind of writing, and when we see it we immediately suspect a cover-up. Professor Richard Mitchell, in his book *Less Than Words Can Say,* analyzes an educator who wrote:

> Our program is designed to enhance the concept of an open-ended learning program with emphasis on a continuum of multi-ethnic academically enriched learning using the identified intellectually gifted child as the agent or director of his own learning.*

Mitchell comments that ''. . . a bewildering exercise in naming may convince the thoughtless and indolent that something must have been told . . . . He [the writer] very probably thinks that he *has* told something and, accordingly, that he has thought something.''

Educators are far from being the only group to indulge in snow jobs of this sort. You are probably familiar with instances in your own field of work. We have found many in ours. Some examples:

Foolish Rhetoric	**Plain Talk**
Despite the inadequacies of the available information, a careful speculation as to the possibilities is necessitated by the importance of discovering observable phenomena of possible significance in petroleum exploration.	This is a poor set of data, but it may contain some information that will help find oil.

---

*Richard Mitchell, *Less Than Words Can Say,* Boston: Little, Brown and Company, (1979), pp. 196–197.

The background information has been surveyed and organizational activities preparatory to initiation of the two component phases of the project being contemplated at this time have been instituted.	After a search of the literature, we have decided to attack the problem in two steps:
Equipment in this category is especially advantageous in that it is easily adaptable for nontechnical personnel in field operations.	Untrained field workers will find this equipment easy to use.
The techniques used in these mineralogical studies were determined to be the most advantageous from both a speed and accuracy standpoint.	We used the fastest, most accurate mineralogical techniques available.
This lack of acceptable performance is explained by the fact that a decrease in the sharpness of the cutting edge renders the blade incapable of penetrating a hard formation.	A dull bit won't cut hard rock.

In a timely effort to make writers everywhere more conscious of the virtues of spare and candid writing, the Document Design Center of the American Institutes for Research, a Washington-based organization, publishes a monthly newsletter called *Simply Stated.** This is a good source of the latest news on the plain-writing movement.

LIMIT AVERAGE SENTENCE LENGTH

For some years writing experts have been admonishing technical writers to prefer short sentences. Some say an average sentence

---

*To get on the mailing list, write Edward Gold, *Simply Stated,* Document Design Center, American Institutes for Research, 1055 Thomas Jefferson Street N.W., Washington DC 20007.

Plain Prose by Law

Spearheaded by New York, more than half a dozen states in the U.S. now have laws that require plain English in consumer contracts. Wisconsin passed a law mandating plain language in all insurance contracts. A manual entitled "Plain Writing in Texas State Agencies"* helps Texas bureaucrats communicate with the citizenry. The National Labor Relations Board has published a style manual to show government judges, lawyers, and rule-writers how to write clear legal prose.

In 1984 a U.S. district court judge, characterizing the language used as "bureaucratic gobbledygook," ordered the Federal Government to rewrite the form letters it was sending to Medicare patients. The Social Security Administration has rewritten its regulations. Even the IRS has been cleaning up its act. Elsewhere, Australia, Canada, Ireland, Nigeria, Romania, Sweden, and the United Kingdom are among nations that are working toward, or already have, plain-language laws on the books.

So far, fears that the replacement of legal terminology would result in a flood of litigation have proved unwarranted. And many businesses—not just computer companies criticized for their "user-unfriendly" manuals—are coming to realize that plain English, in contracts as well as in instructions, can help rather than hurt them.

---

*Irving N. Rothman, "Plain Writing in Texas State Agencies," University of Houston, University Park (1983) 30 pp.

length of 20 words is good; some say 25. Writers respond in different ways. Some continue writing long, complex sentences because they are convinced that short sentences would be "talking down" to intelligent readers. (Is intelligence really insulted by simplicity?) Being afraid of "writing down," these people "write up."

Other on-the-job writers embrace the short-sentence theory readily but overlook both the word *prefer* in "prefer short sentences" and the word *average* in "average sentence length." The result is choppy, primitive, childish writing. In trying to avoid "writing up," these authors do "write down."

An average of 25 to 30 words per sentence is probably about

right for technical writing. This average is only a general guide, of course. A 300-word sentence could be readable if it were punctuated skilfully; semicolons, for instance, have the effect of a period yet relate ideas more closely than a full stop would.

The more complex the concept being described, the more important it is to use short sentences. For example, a complicated piece of equipment should certainly be described in sentences simpler than would be needed for a letter thanking a client for his business. In either case, a two-word sentence can have great impact, while an 80-word sentence, punctuated properly, can provide continuity of thought.

Too Long	Simplified
While a computer program instructs the computer how to do a specific job, other instructions, called the "operating system," which are more basic than the program itself, also reside in the main memory, where they control the program in storing information, printing it, sending it over communication lines (such as telephone lines), and performing many other system functions. (58 words)	A computer program instructs a computer how to do a specific job. Other instructions, more basic than the program, also reside in the main memory; they are called the operating system. This system controls the storing, printing, and transfer (e.g., by telephone) of information, and many other system functions. (Avg. = 16 words per sentence, wps)
This well scratcher is provided with spikes of stiff spring scratcher wire held firmly to the tubing by a rubber-like base, the stiff spring scratcher wire having a portion extending in a direction tangential to the wall of the tubing, and a transversely extending, permanently bent portion embedded in the rubber-like base. (54 words)	This scratcher has spikes of stiff spring scratcher wire firmly attached to the tubing. One end of the wire is permanently bent and embedded in a rubber-like base; the other end extends away from the tubing wall. (Avg. = 19 wps)

In the following table, which shows the results of precipitating with formaldehyde in order to obtain an estimate of the flavonoid and non-flavonoid components in a mixture of catechin (the flavonoid) with vanillic, caffeic, and syringic acids (the non-flavonoids), one observes that the catechin was largely precipitated, whereas the other components of the mixture remained in solution. (57 words)

The following table shows the result of precipitating with formaldehyde to estimate the content of a mixture. The mixture contains catechin as the flavonoid and three acids (vanillic, caffeic, and syringic) as the non-flavonoids. The catechin was largely precipitated; the non-flavonoids were not. (Avg. = 15 wps)

COMBINE CHOPPY SENTENCES

When sentences are too short rather than too long, they can be merged to good effect. Here are a couple of examples.

1. ~~We obtained~~ sixty samples from ~~the~~ Antarctic, an area of particular interest ~~to this work was selected. The area is shown in~~ (Fig. 1). ~~The samples~~ were analyzed ~~by the method~~ as shown in Table I. [**Note:** the unedited version of this paragraph was the product of an author oversold on the short-sentence concept.]

2. The map of the test site ~~is given in~~ (Figure 1). ~~This map~~ shows the 20 scintillometers arranged on the new extended spacing. ~~The location of~~ with the ~~thermometer stations is also shown on the map. These~~ thermometers ~~are~~ placed in the standard pattern.

CUT THE EXCESS VERBIAGE

Redundancies and circumlocutions may still be adding needless words to your text. Get rid of them as in the following examples.

~~We shall appreciate your submitting to our preplacement~~ *You may take your* medical

examination ~~which may be conducted by any reputable~~ *at our* medical

~~doctor. If available~~ *department or* your college health center or college physician

will be ~~very~~ satisfactory *°* ~~for this examination. You should request~~

Ask the physician to complete the attached medical report form and

return it to our Medical Department together with ~~the~~ *his* bill ~~for the~~

~~physician's services.~~

Original: 62 words   Cut to: 36 words

~~The experiments were carried out using~~ *We used* the rolling-ball viscometer

of the Coleman Reservoir Fluid Analyzer. ~~The variation of the~~ *Fig. 1 shows*

low ~~crude~~ viscosity *varied* ~~with~~ mechanical pressure at ~~a temperature of~~

100°F. ~~is shown in the attached Fig. 1.~~ *Fig. 2 shows* The change in ~~the~~ *crude* viscosity

~~of the crude~~ at different concentrations of propane and carbon

dioxide at ~~a pressure of~~ 1,000 psig and ~~a temperature of~~ 100°F. ~~is~~

~~shown in the attached Fig. 2. This latter graph shows that the~~

Crude viscosity of the crude is markedly reduced as the concentration of

propane or carbon dioxide is increased.

Original: 94 words   Cut to: 59 words

~~Some preliminary discussions have been held between representa-~~

~~tives of this company and~~ *We have talked to* the Brandex Chemical Company ~~with~~

~~regard to the possibility of~~ *about* obtaining by-product carbon dioxide

from their plant ~~in the event that~~ *if* the price ~~should prove sufficiently~~

is attractive.

Original: 39 words   Cut to: 21 words

PREFER THE CONCRETE
TO THE ABSTRACT

Authorities Bergen and Cornelia Evans make this point:

The more specific a word is, the more information it conveys.
It is very easy to use words that are too general. In fact, this
is the most obvious characteristic of ineffective writing.[5]

The improvement in the following examples is dramatic.

Abstract, Fuzzy	Concrete, Meaningful
Every precaution was taken to ensure preservation of the samples in a sterile condition.	We took 1,500 pieces of sterilized glassware for sample collection.
Full utilization of the facilities was realized.	The project used six computers.
Advantage was taken of temporary storage facilities.	Laundry hampers served as temporary storage bins.
Owing to the condition of the facilities, inadequate results were obtained.	In two out of three tests the chromatograph malfunctioned and produced no records.

QUANTIFY WHEN POSSIBLE

The operative rule here is: Do not write merely to be understood. Write so that you can**not** be **mis**understood.

Considerable work has been done.	How much is "considerable"?
They appear in limited quantities.	How much is "limited"?
We can save appreciable amounts . . .	$500? $50,000? An estimated 10%? About half? Roughly 60%?
Experience shows that . . .	Whose experience? Ours? Theirs? Everybody's?
Relatively deep drilling . . .	Relative to what? How many feet?
... is significantly greater ...	By 10%? 100%? 1,000%?
... was markedly reduced ...	By how much?
... of the order of magnitude ...	30 $\pm$ 0.2? or $\pm$ 5.0?
It is believed [felt] that . . .	By whom? By us? By people in general?

DELETE UNNECESSARY QUALIFICATION

Qualify when you need to; hedge if you have to. (But hedging has an aura of disrepute. If not done properly, it suggests dishonesty or weakness. Don't waffle excessively or conspicuously.)

If qualifications are called for, by all means put them in, ensuring that they are properly emphasized. If they are not necessary, don't weaken your material through excessive caution. Two guides are helpful: (1) don't qualify any further than you would in conversation (weaseling is much more difficult face-to-face); (2) don't try to put all the qualifiers in one sentence. Fear that a sentence will be taken out of context leads to omnibus sentences so weakened by qualification that the positive facts are lost. The following sentence is a fine example:

Preliminary results may tentatively be interpreted to indicate that an adequately driven bank of the fluid could possibly produce the same ultimate recovery provided that . . . . (eight weakening elements)

We do not have to kill an idea in limiting it. Here are three briefer, adequately limiting ways to qualify the sentence:

1. A driven bank of this fluid could produce the same ultimate recovery if . . .
2. While results are preliminary, it appears that . . . would produce the same ultimate recovery if . . .
3. The results suggest that a driven bank of this fluid would . . . if . . .

EDIT FOR CONTINUITY

One could achieve a good overall structure, write simple sentences with strong verbs, and edit out all the padding yet still leave sections of unclear writing. While the expanded outline ensures that the larger units (chapters and major subsections) will hang together in a clearly understood pattern, the pattern is harder to maintain at the paragraph and sentence level, where an outline offers no guidance.

As a result, the writer is apt to leave gaps in logic that the reader's experience cannot bridge. When the discussion changes course without warning—reverses, jumps to a parallel track, skips too far ahead, digresses briefly—the reader has to back up a few paragraphs and concentrate. If the writer has a pattern in his mind, he must transfer it systematically to paper.

Use Signal Words

Signals are transitional words meaning "Go," "Slow Down," "Reverse," "Switch," etc. Note in the elliptical paragraph below how the signal words guide you:

Besides these qualifications, it is imperative that ....GO
It is important, moreover, that.............................GO
On the other hand, it does not matter that......... REVERSE
Notwithstanding these exceptions,.........REVERSE AGAIN
So it becomes evident that .... GO TO END

The signal words most commonly used in technical writing are *also, in addition,* and *however.* As suggested by Fig. 23, many others enrich the language. Learn to use a variety of them.

Obviously, too many of these signals would slow the pace of writing and become monotonous. We therefore use other signalling techniques as well.

Use the Pick-Up Technique

When a gap is obvious, or when the transition from one thought to another seems abrupt, try to pick up a word from one paragraph or sentence and carry it into the next, using a now-familiar term to lead into the new idea.

Here is the last sentence of one paragraph, for example, with an abrupt change to the first sentence of the next paragraph:

. . . Therefore few organizations have the necessary technology for this kind of analysis.

Not even extensive familiarity with principles and equipment can take the place of experience, through which . . .

The switch here is from technology to experience. There is no warning of the switch and no close tie between the two. But we can pick up the word organization and use it as the tie:

. . . Therefore few organizations have the technology necessary for this kind of analysis.

PARALLEL TRACK SIGNALS

TO INTENSIFY:
As a matter of fact,
In any case,
In any event,
Indeed,
In fact,
Obviously,
That is,

TO DENOTE TIME:
At the same time,
In the meantime,
Meantime,
Meanwhile,
Simultaneously,

TO REPEAT:
As I have said,
As noted above,
In brief,
In short,
In other words,
In summary,

TO GIVE EXAMPLES:
As an illustration,
For example,
For instance,
To illustrate,
To demonstrate,

TO COMPARE:
By the same token,
Correspondingly
Equally
Equally important,
In the same way,
In the same manner,
Likewise,
Similarly,

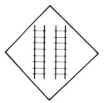

GO SIGNALS

TO DENOTE TIME:
After a few hours,
Afterward,
And then,
At length,
Finally,
First, second, etc.
Formerly,
Immediately
    thereafter,
Later,
Next,
Previously,
Soon,
Subsequently
Then,
Thereafter,
While

**TO SHOW
PURPOSE:**
For this purpose,
For that reason,
In order to
So
To
To this end,
With this object,

TO SHOW ADDITION:
Additionally,
Again,
Also,
And,
And then,
Besides
Equally important,
Finally,
First, second, etc.
Further,
Furthermore,
Last,
Moreover,
Nor
Too,
What's more,

CHANGE DIRECTION SIGNALS

TO MAKE CONTRAST:

Although	On the other hand,
Although it is true,	Still
At the same time,	Though
Be that as it may,	Yet, And yet,
But	
Conversely,	
Even so,	
For all that	
However,	
In contrast,	
Nevertheless,	
Nonetheless,	
Notwithstanding	
On the contrary,	

TO REVERSE:
Going back to
Referring back to
Regressing to
Returning to
Reverting to

Figure 23a   Some signal words that alert readers

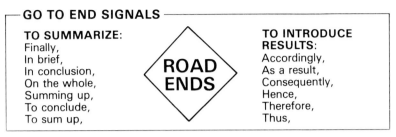

TO SUMMARIZE:
Finally,
In brief,
In conclusion,
On the whole,
Summing up,
To conclude,
To sum up,

ROAD
ENDS

TO INTRODUCE
RESULTS:
Accordingly,
As a result,
Consequently,
Hence,
Therefore,
Thus,

Figure 23b   More signal words

Still fewer organizations have the experience needed. Not even extensive familiarity with principles and equipment can take the place of experience, through which . . .

In the following paragraphs, note the abrupt stops and starts:

No Signals

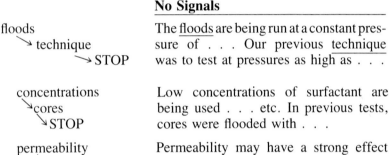

floods
→ technique
→ STOP

The floods are being run at a constant pressure of . . . Our previous technique was to test at pressures as high as . . .

concentrations
cores
STOP

Low concentrations of surfactant are being used . . . etc. In previous tests, cores were flooded with . . .

permeability

cores

Permeability may have a strong effect on test results. The rocks now under study have permeabilities of . . ., whereas in previous tests the cores had permeabilities ranging from . . .

By picking up a word in each paragraph and carrying it into the next, we can proceed smoothly from the old into new topics:

Revision 1—Signals Added

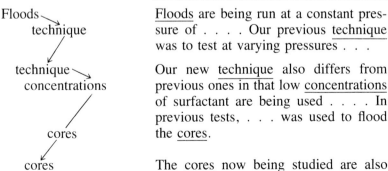

Floods
technique

Floods are being run at a constant pressure of . . . . Our previous technique was to test at varying pressures . . .

technique
concentrations

Our new technique also differs from previous ones in that low concentrations of surfactant are being used . . . . In previous tests, . . . was used to flood the cores.

cores

cores
permeabilities

The cores now being studied are also different from those previously flooded: they have much higher permeabilities.

List to Show Relationships

You may already have observed that this example is perfect for another kind of signalling—a numbered or verbal list.

Revision 2

<u>three</u> = a signal   The new technique differs from previous methods of flooding in <u>three</u> ways:

(1) GO   (1) Floods are being run as a constant pressure . . .

(2) KEEP GOING   (2) Lower concentrations are . . .

(3) STOP   (3) Cores of higher permeability . . .

Revision 3

<u>three</u> = a signal   The new technique differs from previous methods of flooding in <u>three</u> ways. <u>First</u>, these floods are being run

<u>First</u>   GO

<u>Second</u> KEEP GOING   at a constant pressure of . . . <u>Second</u>, we are using lower concentrations of surfactant . . . . <u>Finally</u>, the cores

<u>Finally</u> STOP   have higher permeabilities . . .

LINK IDEAS TO SHOW EQUAL RANK (PARALLELISM)

Use the coordinating conjunctions *and, or, nor, for,* and *but* to show that ideas or groups of words are of equal rank. These conjunctions (and pairs, such as *either-or, neither-nor*) are intended only for elements that are alike in construction, feeling, or content.

and \\\clause \\\clause   **but** \\\phrase \\\phrase

or \\\infinitive \\\infinitive   **and** \\\idea \\\idea of equal rank

In polished writing, conjunctions that show equal rank are not used to join unlike elements:

(1) **This:** The data will be used <u>to</u> <u>analyze</u> profitability and <u>to</u> <u>learn</u> more about last year's disaster.

. . . will be used   \\to analyze
                        and\\ to learn

Not this: The data will be used to <u>analyze</u> profitability and
in <u>learning</u> more about last year's disaster.

. . . will be used   \ / to analyze
                        and ✕ in learning

(2) **This:** The mixture became <u>both thicker and browner</u>.

The mixture became **both**   \\thicker
                          and\\ browner.

Not this: The mixture both became thicker and browner.

The mixture **both** \ ✕ / became thicker
                and / ✕ \ browner.

To prevent awkwardness, make the clauses similar in feeling
as well as structure.

(3) **This:**

\\ Fig. 1 shows the area from which samples
\\   were collected and analyzed,
and \\Fig. 2 shows the results of sample analyses.

Not this:

\ / Fig. 1 shows the sampling area
and ✕ we have analyzed samples from this area.
/ \ The results appear in Fig. 2.

In technical writing, correct subordination of ideas may be
more critical than parallelism in revealing pattern, showing relationship, creating emphasis, and indeed, in expressing meaning.

Examine the following two sentences, which are not shown to
be related in any manner.

It appears that the differences in compounds A and B are
due to different manufacturing techniques. A spectrometric
analysis is in progress.

Now observe how subordination of one of the ideas adds to or changes the meaning:

(1) <u>Although</u> it appears that the differences in compounds A <u>and B</u> are due to different manufacturing techniques, a spectrometric analysis is in progress.

(2) <u>Because</u> it appears that the differences . . . a spectro-metric analysis is in progress.

In English, the variety of linking words and phrases that show one idea is dependent upon the other gives us great flexibility. Here are common ones:

after	in order that	that
after which	if	to which
all of which	inasmuch as	though
although	in that	unless
as	in which	until
as if	no matter	when
as soon as	none of which	where
as though	none of whom	whereas
at which	prior to	while
because	since	which
by which	so long as	who
even though	so that	whom
for which	than	whose

EDIT FOR VARIETY

Variety at the sentence level makes for pleasant reading and keeps the reader interested. Here are two good ways to achieve it.

Vary Sentence Structure

A vice-president in a large organization once gave the staff editors some of his in-house articles for analysis. He didn't know what was wrong with his style, but it bored him. Apparently it bored his boss, too, he said, for nearly everything he composed came out of the president's office revised.

Examination of the VP's articles showed that he wrote simply; his meanings were clear; his vocabulary was excellent; the material was flawlessly organized. According to all the readability tests published, his articles should have been highly readable. They were understandable, but they were also—as the VP suspected—extremely boring.

The problem was monotony, due only to the fact that his sentences were nearly all of a single type. In a 40-sentence article, he used 39 sentences classified as "simple" (made up of a single independent clause) and one classified as "complex" (an independent clause with one dependent clause).

The importance of varied sentence structure can be grasped quickly if you will first read aloud the steps in this diagram, which represents a sample of the monotonous, equally measured steps of that VP's writing: chop-chop; chop-chop; chop-chop; etc.

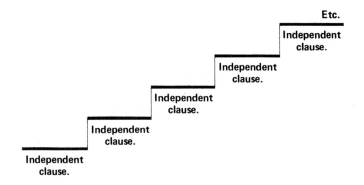

Enough of that could lull you to sleep. Now contrast that with the rise and fall of your voice as you read aloud the "steps" in an article whose sentence types vary pleasantly:

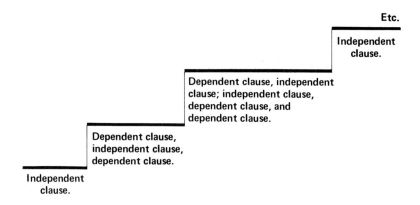

Four suggestions on sentence structure:

(1) Where ideas are of equal rank, occasionally combine simple ones to make compound sentences:

Fig. 1 shows the sampling area, and Table I summarizes the analyses.

(2) Where one idea is dependent upon another, show this relationship by combining them in a complex sentence:

Since the Gulf area is of particular interest, six samples were obtained from the area and analyzed.

(3) Where a number of ideas are interrelated, use a compound-complex sentence (at least two main clauses and at least one dependent idea):

Since the Gulf area is of particular interest in these studies, six samples from the area (Fig. 1) were analyzed and the values were calculated as shown in Table I.

(4) To avoid undue complexity, use simple sentences frequently. If you habitually write complicated sentences, your problem will be making use of the simple construction. But in any case, **strive for a nice balance.** A report having only simple sentences is boring. So is one having only long, complex sentences.

Vary Placement of Clauses for Variety

In constructing the more complicated kinds of sentences, don't become addicted to a particular placement of dependent clauses—at the beginning of the sentence, for example. Here are three ways of placing one dependent idea in an example sentence:

(1) [Since the Gulf is of particular interest,] we have analyzed six samples from this area.

(2) In the Gulf area, [which is of particular interest,] we obtained six samples for analysis.

(3) We have analyzed six samples from the Gulf area, [which is of particular interest.]

Sometimes it is important (especially if you are editing somebody else's material) to notice that there may not be a choice in placement of clauses. A particular construction may be required to emphasize a particular idea.

Sentence #1 below emphasizes tests, while Sentence #2 emphasizes the buildup:

(1) The **tests proved** that accumulations do build up rapidly.

(2) That **accumulations build up rapidly** was proved by the tests.

Placement also influences continuity. The need to achieve a smooth transition between a sentence and a following or preceding idea often dictates the construction to be used.

Vary Sentence Length for Variety

Here is a paragraph with a monotonous sentence pattern:

47 words, 11 + words per sentence (wps):

The fluid had been used for some years before these tests. Its users believed the fluid to be an extraordinarily cation-sensitive one. Therefore it was their feeling that it could be handled and circulated with advantage. This was recently demonstrated to be true at this laboratory.

Here is the same information with the same average sentence length but with a more interesting pattern (and please note that its length has been cut):

35 words, 11 + wps:

For some years chemists have used the fluid in the belief that it is extraordinarily cation-sensitive and thus can be handled and circulated with advantage. It can. This was recently demonstrated in our laboratory.

In diagram, the patterns of these two paragraphs clearly demonstrate the advantage of varying sentence length:

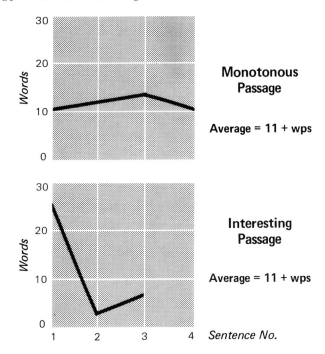

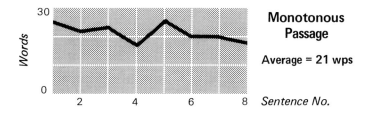

The next two diagrams contrast "pattern interest" more forcefully because they represent a longer article. This passage was originally written in a range between 17 and 24 words per sentence, with an average sentence length of 21 words:

The passage was rewritten in a range between 3 and 75 words per sentence. The average is still 21, but see how much more interesting the pattern is:

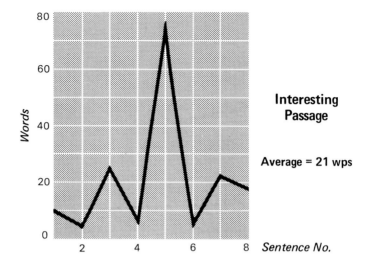

The difference between these patterns is the same as that between two rooms, one decorated in shades of gray and the other
with sharp accents of bright color. The grays represent the majority of sentences—not very long, not very short. The accent colors
represent the very long and very short sentences that prevent monotony.

A FINAL CHECK:
PICK THE SKELETON

As you edited for good organization, you undoubtedly marked
some poor sentences with a "?" or an "X," meaning "Something
is wrong here; it's awkward; it isn't clear." Writers often work on
such passages at length, trying one rewrite after another without
satisfaction. The trouble is they haven't identified what's wrong
with the sentence and therefore don't know what they're fixing.

An easy approach to unruly syntax is to do a simple sentence
analysis that we call "picking the skeleton." Skim through a sentence and pick out the subject and verb, or the subject, verb, and
complement (the word or idea that completes the thought). See
what this skeleton says. Then look at the modifiers.

If this smacks too much of grammar, don't be alarmed. You
need not be a grammarian to use a simplified version of the technique.

A Substitute Analysis

If you were to compress a full sentence into an informative side heading, you would get pretty close to a grammatical sentence skeleton. Suppose, for example, that the NEWS in your monthly progress report is "Data collection has been delayed because of higher-priority work." A side heading you would write for that NEWS might read, "Data Collection Delayed." That is close to the grammatical skeleton (subject and verb): "Collection has been delayed."

That example is too simple to need analysis, of course. The next sentence is neither simple nor clear, and the author had trouble rewriting it.

Slight changes were accomplished, but it points out that variations in fluorescence can be encountered when the additive is altered with respect to concentration.

To analyze, or diagnose, the problem here, the author and editor took the sentence apart. They identified four ideas:

(1) changes were accomplished

(2) it points out

(3) variations can be encountered

(4) additive is altered

Here's what we learn from those four skeletons:

(1) "Changes were accomplished" contains a weak verb that is one of the classic danger signals of a distortion. Something changed?

(2) In "it points out," what does it refer to? Not changes, for a singular pronoun cannot refer to a plural noun. (Faulty pronoun reference.)

(3) In "variations can be encountered," the verb can be encountered is weak, and the subject is one of those -tion nouns that could be a distorted verb. Does something vary? Is that the "slight change" mentioned? Does fluorescence vary slightly? Or—

(4) In "when additive is altered," is that alteration the "slight change"? Or were both the variation and the alteration slight?

"No," the author said, "it wasn't the additive that was altered, actually; we changed the concentration of the additive. What I mean is:

"Fluorescence varies with slight changes in the concentration of the additive."

Table 4 shows how easily sick sentences can be diagnosed by picking the skeleton, so long as the sentences are relatively uncomplicated.

TABLE 4
Picking the Skeleton to Diagnose Sick Sentences

Complete Sentence	Skeleton	Diagnosis
1. (a) Delays in the shipments referred to in your letter of Jan. 6 have occurred due to a strike in our Chicago plant.	delays have occurred	Uninformative skeleton; weak verb; misuse of adjective "due" as a verb modifier.
(b) The shipments referred to in your letter of Jan. 6 have been delayed by a strike in our Chicago plant.	shipments have been delayed	Good skeleton; healthy sentence.
2. (a) Studies of a new method of classifying oils have been successfully completed.	studies have been completed	Lab-centered; author-centered. Unless the reader is an impatient supervisor, who cares? *Successfully* modifies *have been completed*. Was it the completion that was successful or the method?
(b) An accurate method of classifying oils, developed in our recent geochemical studies, groups the oils according to their source.	method groups oils	Skeleton gives information about the method, not the study. What the author did is acknowledged but subordinated. Good skeleton.

3. (a) Reduction in flow rates are accomplished as a result of this change.	Reduction are accomplished	Subject is singular, verb plural. Distorted verb (reduction); weak verb (are accomplished); wordy, backward.
(b) This change reduced flow rates.	change reduced rates	Gets information in the skeleton. Sentence is short, as it should be.
4. (a) By constructing this corridor, the animals can move freely from cage to cage.	animals can move	Good skeleton, but the opening phrase dangles. There is only one noun the phrase can modify—animals—and they cannot construct a corridor.
(b) This corridor allows the animals to move freely from cage to cage.	corridor allows animals to move	This is more sensible.
5. (a) Modification of the equipment is anticipated before initiation of a field test is attempted.	modification is anticipated initiation is attempted	Distorted verb (modification); foolish verb (is anticipated). Field test must be dangerous if one has to think about even attempting to start (initiate) it!
(b) We will modify the equipment before field-testing it.	we will modify equipment	A simple idea deserves a simple expression. Skeleton makes sense.
6. (a) Inferences may be extended to a reasonable position from which a decision may be made as to the advisability of collecting additional information.	Inferences may be extended decision may be made	A trivial idea "extended" beyond a "reasonable position." In a really garbled sentence like this, you have to reach outside the skeleton to find some indication of what the idea is.
(b) If this procedure does not solve the problem, we will need to collect more data.	procedure does [not] solve problem we will need to collect data	This is one reader's best effort to interpret a very "sick" sentence.
7. (a) The studies were conducted in Sand Springs, where the subsurface is relatively uniform.	studies were conducted subsurface is uniform	Skeletons are OK, but the author noted that the modifying clause "where the subsurface is relatively uniform" does not describe all of Sand Springs; it describes that part where studies were conducted.
(b) The studies were conducted where the subsurface is relatively uniform in Sand Springs.	same skeletons—	—but misplaced modifying clause has been moved next to the word it describes.

Sentence Analysis: The Real Thing

Bear in mind that the above method of picking the skeleton is only a substitute for a standard grammatical analysis. It may fail if you cannot distinguish between active and passive verbs or between verbs and verb look-alikes (participles, gerunds, verbal nouns, and infinitives).

The following verb tests should tell you whether you are capable of diagnosing sentences that contain those elements. If you fail the tests, you may need to refresh your memory with a grammar review like that in House and Harman's *Descriptive English Grammar*[3] or the *Harbrace College Handbook*.[4]

Test 2

IDENTIFYING VERBS

If you needed to determine whether the verbs in the following sentences are used efficiently (active or passive), could you begin by telling which are the verbs and which are other parts of speech?

Directions:

In the following sentences, underline the verbs. Do **not** underline infinitives or forms derived from verbs (participles, gerunds, verbal nouns). Answers are given on p. 259.

(1) Improvement of the 25-foot-wide dock included the installation of a new fendering system.

(2) Now the largest ships anticipated can be accommodated without overstressing the dock.

(3) This system extends the full length of the 1,208-foot-long dock, long enough to berth two 745-foot vessels simultaneously.

(4) New pile clusters, one upstream and the other downstream, are used for mooring lines.

(5) Storage and subsequent delivery, however, can present problems, especially in very cold weather.

(6) To provide storage, six 72,000-bbl tanks were converted by adding steam coils and insulation.

(7) All are cone-bottom tanks so that they can be completely emptied.

(8) Coils in each tank total approximately 6,500 feet of 4-inch pipe.

(9) A Varek tank-gauging system is used to keep the terminal operator informed of the contents and temperature of each tank.

(10) The boilers produce a substantial volume of 15-lb maximum steam, which is circulated through the tanks, and which maintains a temperature of 115° to 155°F.

Test 3

PICKING THE SKELETON

To determine whether a verb is active or passive, you should be able to recognize subjects, verbs, and objects.

Directions:

(a) In the following sentences, underline the subjects once, the verbs twice, and the complements three times. (The complement is the word, phrase, or clause needed to complete the skeletal idea sensibly.)

(b) If the complement is a dependent clause, instead of underlining the whole clause three times, put brackets around it. Then underline the skeletal elements *within* it.

Example:

This report does not indicate [that new pile clusters are needed].

(1) The head produced by a particular pump is the height reached by the pumped liquid.

(2) The head of a pump is normally given in feet.

(3) From the topic on pressure, you will remember that there is a relationship between the height of a column of liquid, its specific gravity, and the pressure exerted at the bottom of the column.

(4) Therefore, the head produced by a particular pump under certain operating conditions is an indication of the pressure (psig) developed by the pump.

(5) The location of the pump and the level of the liquid to be pumped determine the maximum height to be reached by the liquid.

(6) To fully understand this subject of head, look at the following pumping systems.

(7) Assume in all these systems that the pumping is being done by the same pump, and that the liquid being pumped is water.

(8) Note in Figure 2 that water rises to the same level in the tank and pump.

(9) In this case pumping is raising the liquid 100 feet.

(10) Therefore we say that the discharge head developed by the pump at a capacity of 100 gpm is 100 feet.

Test 4

DISTINGUISHING ACTIVE FROM PASSIVE VERBS

The chief clear-writing technique is concerned with the voice of the verb. Can you tell whether verbs are active or passive?

Directions:
In the following sentences, underline the verbs and state whether they are in the active (A) or passive (P) voice. If the verb has no voice, mark it intransitive (I).

(1) Business leaders could hardly believe their good fortune.

(2) They were offered a facility near the downtown hub.

(3) Headed by Julius Jamieson, later a justice of the State Supreme Court, a committee studied the advisability of accepting the gift.

(4) It was suggested that the name of the facility be changed to honor him.

(5) A century after the name was selected, the choice had unexpected consequences.

(6) Halley's comet is sometimes called the "once-in-a-lifetime" comet.

(7) The lifespans of relatively few people coincide with two passes of the comet around the sun.

(8) Halley's comet was expected to be visible through binoculars and amateur telescopes.

(9) The comet was visible from Earth in 1910.

(10) It is said that frightened people bought comet insurance and comet pills in case something terrible happened.

Test 5

VERBS AND VERBALS

How much do you remember about verbals (participles, gerunds, and infinitives)? As technical writers tend to convert good strong verbs into these weaker forms, you need to be able to recognize them. You should also recognize when a group of words does not form a complete sentence.

Directions:

In the following eight units mark the complete verbs with a double underline. Mark any verbals (gerunds, participles, and infinitives) with a single underline. Finally, mark the whole unit either **complete** (for a complete sentence) or **incomplete.**

(1) Constant lobbying to prevent damaging legislation is credited with influencing disinterested legislators.

(2) In evaluating a merger, the fact that an industry is ailing or a company is failing.

(3) Discovery of the genetic code grew out of having a theory different from accepted wisdom.

(4) More hiring in companies producing new products and expanding existing lines and thus boosting employment.

(5) The greatest problem will be developing a reasonable explanation for this phenomenon.

(6) Not buying insurance to cover medical costs that cannot be predicted and that are unlikely to be covered adequately.

(7) Managing your money is the subject of the next lecture.

(8) Before proceeding with the next step, to enclose the sample in a coating of plastic.

Edit for Correct Grammar and Current Usage

This section is designed for writers who wish to use correct English but who may be confused about present-day standards as well as certain isolated points of grammar and usage.

WHAT IS CORRECT ENGLISH IN NORTH AMERICA?

A short answer to the question is Standard American English— the constructions preferred by the majority of educated people in North America. To be preferred, a grammatical construction should be supported by logic or by continued use by reputable writers—preferably both.

That is a good definition, but as the authorities listed in our references demonstrate, educated people and reputable writers do not speak with one voice. We get conflicting advice as to what is currently acceptable, depending on the linguistic philosophy of the advisor. Some authorities are permissive, some are tyrannical, and some are reasonable defenders of high standards. This leaves us with options to choose from and decisions to make.

Permissiveness or Laxity?

Certain influential and highly vocal linguists have argued against rules for correctness. The authority of grammarians is suspect, they say, and grammar rules are arbitrary instruments of class prejudice; change is inevitable, and it is absurd to label departure from rules as good or bad; social changes in a pluralistic society lead to new grammars that are as logical as those of the "educated middle class"; etc. So what right does a grammarian or dictionary have to set standards? That is what these linguists say.

Webster's lexicographers agreed with this view, and in 1961 the *Third International Dictionary* appeared with few status labels or usage notes. Those included were permissive. For instance, *ain't* was treated this way:

> **ain't**   though disapproved by many and more common in less educated speech, used orally in most parts of the U.S. by many cultivated speakers, esp. in the phrase *ain't I?*

The good book many of us needed when in doubt—or in an argument—thus became a mere list of words in use. A public uproar ensued, and ten years later the *Living Webster Encyclopedic Dictionary* revised this entry:

ain't . . . is so widely regarded as a mark of illiteracy that it should generally be avoided. *Ain't* is never acceptable in written English.

By the time the newer dictionary was distributed, the educational establishment had dedicated itself to permissiveness. Many college graduates, including some teachers, entered their professions unable to meet the minimum language standards of their employment.

Moreover, the problem has been heightened for students by constant exposure to the ethnic, rock-music, drug-culture, and underworld jargons popularized on campus. As a result, many young professionals make formal presentations laced with slang and punctuated with *like, y'know, basically,* and *in terms of* as constant fillers between thoughts for which ordinary meaningful words seem hard to find.

The communications media have had a more pervasive influence on the way we write, talk, and think than we are conscious of. Advertisers so constantly assault our eyes and ears with wretched language that our sensitivities are bound to be dulled. Some examples:

You ain't got a thing if you ain't got that cling.

I don't wear nothin' at all.

A company makes their own cheeses themselves.

The soup that eats like a meal.

It's a moot question whether Madison Avenue copywriters regard their sales targets as illiterates, whether they just want to insult us so outrageously that we can't forget the product, or whether certain copywriters are themselves close to illiterate.

A professor on a televised college lecture program said, "Prices begun to fall . . . and Americans begun to demand more protein." The editor of a journal for technical writers allowed this caption: UH Discusses Their Masters of Technical Writing. A book on technical writing spells "grammar" as "grammer" not once, but many times.

Even music contributes to poor language habits. Rock, western, and country lyrics are composed largely of substandard English. "Authenticity" may excuse it, but constant exposure affects us. Children do their homework by it. Adults tune it in as they dress for the day, drive to work, jog through the park.

It would be easy to blame mere permissiveness for the depressing speed with which these noxious influences have spread. Some effects are clearly the result of simple ignorance, but others seem bad enough to qualify as sabotage. Whatever the cause, we need to remain on guard in order to avoid drifting, unaware, into the usage of illiterates.

Restrictive Influences: The Other Extreme

Some grammarians, often called purists, regard change in language as synonymous with decline of language. These peddle a certain amount of nonsense and call it grammar; they live by every rule formulated during the 18th century.

Before 1600 there was no English grammar to study; the word *grammar* referred to Latin grammar. Thus the grammarians of the 17th and 18th centuries understandably patterned many of their new English rules after the Latin rules. They also applied their own logic arbitrarily to the living language, creating rules to govern, for example, double negatives and *shall* and *will*.

For the most part the work of those early grammarians was helpful. Nevertheless, a few pieces of nonsense begun in that fashion still plague us, even though continued scientific analysis of English should long ago have wiped them out. Writers still worry about them: Can one end a sentence with a preposition? Split an infinitive? How about *shall* and *will?* (These and other problems, like dangling participles and *data is* vs *data are,* are discussed under "Questions On-the-Job Writers Frequently Ask," pp. 181–190.)

Champions of High Standards

Grammarians are nevertheless the ultimate authority for technical, scientific, and business writers. They try to preserve a system for communicating in a way that can be understood—that cannot be misunderstood. Our best authorities are contemporary grammarians who recognize changes in usage.

In addition to textbook grammars, we have counsel against declining standards from such careful writers as Henry Fowler, Bergen Evans, Theodore Bernstein, Jacques Barzun, Wilson Follett, S.I. Hayakawa, Richard Mitchell, William Safire, Edwin Newman, and William Strunk Jr. and E.B. White. However, they do not all agree in all matters (e.g., should *contact* be accepted as a verb? *hopefully* as a sentence modifier?).

Where Do Professional People Stand?

We should be the comfortable masters of rules, not vice versa. For our purposes we can disregard the arguments of linguists. The arguments are interesting, but linguists appear to write **to** each other and **at** grammarians, not **for** us. We can likewise forget puritanical grammarians who resist change of any sort. Obviously language changes: Old English is unintelligible, Shakespearean language is dead, Victorian rules are outdated, and arbitrary restrictions are discarded as the need for new terms arises.

This is not an invitation to exhibitionism, of course; nor is it an invitation to take liberties with standards. You can depend on this: you will be judged unfavorably by readers, listeners, employers, and most of your peers if you use nonstandard or substandard grammar, or if you flout contemporary standards of good usage.

The safest course for technical, business, and scientific writers is (1) to use grammar as a basic aid to clear communication; (2) to avoid those usages that classify one in a way that limits career development; and (3) to pursue the ideal of excellence as demonstrated by reputable writers.

QUESTIONS ON-THE-JOB WRITERS FREQUENTLY ASK

The questions asked by writers often concern minor matters of grammar, usage, spelling, and punctuation. However minor the questions seem, not knowing the correct answers invites major problems. Readers who detect bad grammar also tend to downgrade content. Further, mistakes in grammar, usage, punctuation, and spelling are usually caught somewhere in up-the-line editing, and that means retyping pages and losing time.

1. Can We End a Sentence with a Preposition?

Yes—because a preposition is sometimes the best word to end a sentence **with.** The Latin preposition was literally a prepositioned word; in accordance with its name and use, it had to come *before* something. That's how we came by the "rule" that you can't end a sentence with a preposition. Nevertheless, modern English includes a long list of such constructions:

What is that all **about?**

This is a manufacturer (whom) we can count **on.**

I don't know what it is **for.**

That's a deep subject (which) we're getting **into.**

People who tolerate pedantry are responsible for the kind of foolishness that once appeared on an office bulletin board. The company credit union was publicizing money to lend; below a breezy picture of a hand in which a real penny was taped was this stiffly "correct" note:

"—And there's more from where this came!"

2. Can We Split Infinitives?

Sometimes we can; once in a while we need to. Splitting the *to* from the verb can cause awkward constructions. But it is permissible **to** occasionally, when the situation warrants—as it does not in this example you are reading—**split.** In fact, in certain situations, splitting an infinitive can be the only way to write without ambiguity. For instance:

(a) The records completely fail **to define** the species. (Do not define it at all.)

(b) The records fail completely **to define** the species. (Do not define it at all.)

(c) The records fail **to define** the species completely. (Do not define it at all? Or define it only partly?)

To express the author's meaning clearly—which is that the records do define the species, but only partly—we must split the infinitive:

The records fail **to** completely **define** the species.

3. Is the Dangling Participle Really Evil?

Yes, it is chiefly evil. While some dangling constructions are essentially inoffensive, many are like a distorted picture on the television screen—they interfere with reception even though they may not be technically identified. Moreover, quite a few are ludicrous:

> Following this procedure, the photographs revealed no anomalies.

"Following" is a bad dangling participle. Somebody or something not named in the sentence had to do the following. It couldn't be the photographs, and it couldn't be the anomalies that followed the procedure.

> *Revised:* The photographs taken by this procedure revealed no anomalies.

> Participating in these training sessions, a clear picture of management policies can be gained. (Who does the participating? picture? management? policies?)

> *Revised:* Participating in these training sessions will give you a clear picture of management policies.

Clearly these sentences need revision. But not all opening participles are danglers. In English, we often use independent participles:

(a) Speaking of photographic procedures, has the Company decided what procedure to use in Alaska? (Speaking does not modify any word in the sentence. It is idiomatic, and its use as an independent element is considered acceptable.)

(b) Considering all these factors, it is not surprising that the program ran several hours overtime.

Distinguishing between the dangling participle and the independent participle can be hard for ESL writers to figure out. (If in doubt, consult a grammar book or a knowledgeable person.) Writers using English as a native language should be able to make the distinction.

We must also point out that participles are not the only elements capable of being dangled. Almost any modifier can hang loose.

> **Phrase:** As a valued Monnybank customer, we've increased the credit limit on your account. (We [the bank] are the valued customer? That's the only human reference in the sentence that could be a customer.)
>
> **Clause:** Although only a few months old when it started leaking, the tank's coating was deeply pitted and scored. (Obviously it was the tank that started leaking, but tank does not appear in the sentence. Tank's is a possessive modifier and could not leak.)

4. Data Is or Data Are?

Webster's second edition of the unabridged dictionary (1957) gave this usage note: "Although plural in form, data is not infrequently used as a singular; as, *this data has been furnished for study and decision.*"

As one might have expected, the trend toward acceptance of "data is" has continued. What we would not have expected is that, as this book goes to press, the usage is still controversial in technical and scientific circles.

The *American Heritage Dictionary* (1979) has this usage note: *"Data* is now used both as a plural and as a singular collective. *These data are inconclusive. This data is inconclusive.* The plural construction is the more appropriate in formal usage. The singular is acceptable to 50 per cent of the Usage Panel."

The Usage Panel probably reflects the attitudes of readers quite accurately. If we use *data is,* 50 percent of our readers will approve, but 50 percent will criticize. If we use *data are,* 100 percent of our readers will approve (although 50 percent may not care).

Take your choice. We prefer the usage less likely to invite criticism, but when we want to make the noun collective (as the body of the data rather than its pieces), we do not hesitate to make *data* singular.

5. Who or Whom? Whoever or Whomever?

As Bernstein points out in *The Careful Writer,*[9] linguists (and, we might add, permissive English teachers) have tried for a long time, without complete success, to bury *whom*. In formal writing "careful writers" will not tolerate the solecisms that result from misuse of these pronouns.

It takes a little knowledge of sentence structure to understand when to use *who* and *whom,* and that probably accounts for the attempt to get rid of *whom*.

Two rules are simple. Use *whom*—

(a) As the direct object of a verb:

Whom did you elect?

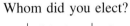

He is a man whom all trust.

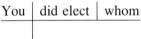

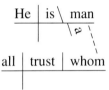

To whom it may concern

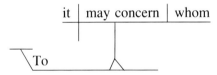

(b) As the object of any preposition (*for, to, with,* etc.):

the person for whom it was intended

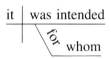

the people with whom we associate

people in whom we have faith

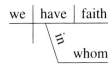

It takes a bit more knowledge to use *whom* correctly in sentences containing a whole idea (clause) as the object of a verb. Within that clause, the pronoun may be the subject, as in the following:

(a) We will hire whoever [not *whomever*] qualifies.

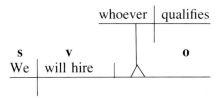

(b) I interviewed several applicants who [not *whom*] I think would qualify.

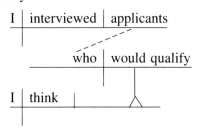

On the other hand, the pronoun may introduce a clause yet serve as the object of the verb:

(c) This is the applicant whom I interviewed.

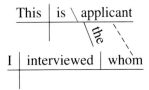

6. What About *Shall* and *Will*?

Although the British by and large still try to follow the old rules for *shall* and *should,* and *will* and *would* (you might remember this in writing for a British journal), Americans who speak standard English tend to ignore them. We do use *shall* in legal documents to express obligation.

It now appears to be acceptable, especially in letters, to write "I will be glad" or "We will forward the extra copies," etc. As Bergen Evans says, "The American people, etymologically speaking, are a *will*ful people." And Fowler, the authority for British usage, appears ready to give up, commenting, ". . . it seems clear that attempts to repel this particular invasion from the other side of the Atlantic have . . . [a negligible] chance of success . . . ."

7. Are Contractions Permissible?

It depends on the degree of formality that is appropriate. Contractions would sound a bit breezy in a formal document such as a scientific report or a formal letter to a head of state. Letters, memos, internal reports, and personal publications can be as formal or informal as you choose to make them, depending on the readers you have analyzed.

8. What Can I Do About My Poor Spelling?

In his book *On Language,* William Safire accuses, "Most of us are proud of two things we ought to be ashamed of: illegible handwriting and poor spelling."

Poor handwriting we could improve with more care and less haste. Poor spelling is a different matter, a harder task. Our system of spelling is full of apparent inconsistencies. Furthermore, we all see words differently, depending on the speed at which we read, the extent of the eye's span, and the way we were taught to read. One who had cards flashed at him in primary grades and learned to recognize words by their shape instead of their component letters does not have the practice in sounding out words that makes spelling an easier, more logical affair.

Nevertheless, improvement is worth working at, for our society tends to equate bad spelling with illiteracy. We suggest the following:

—Use a spelling checker with your word processor.

Lacking a word processor—

—Develop an awareness of the kinds of spelling errors you make.

Do the words ending in *-ence* or *-ance* bother you? Words ending in *-ible* or *-able?* *-er* or *-or?* Do you wonder whether it's *ei* or *ie? -cede* or *-ceed? -ly* or *-ally?* When to double consonants?

—Make a classified list of the words you consistently misspell or have to look up.

You know many of them. Other sources for your list are the editorial corrections made by your supervisors, your secretary, or your student advisor or instructor. Tack the list on the wall, add to it as necessary, and refer to it frequently.

—Consult the dictionary.

This should perhaps have been the first suggestion, for it is an obvious answer to the problem—so obvious that most writers do it first. The trouble is that really poor spellers can't spell well enough to find the word in the dictionary. In that case, depend on a good secretary if you are lucky enough to have one. (That's one of the advantages of dictating.) Check your secretary's or typist's spelling against the dictionary for a while; if it beats yours, depend on it.

9. Can Signs and Symbols Be Used in Text?

If an expression is cumbersome, use the signs and symbols: 1¼″ ID by 1⅜″ OD; 28′ × 11′ × 3″. Otherwise, write them out: feet, inches, percent, etc.

10. Are There Any Simple Rules for Punctuation?

All the formal rules for punctuation are simple individually, but there are a lot of them to remember. (Our "Dictionary of Common Problems," which follows this section, includes rules for the use of commas, colons, semicolons, and hyphens.)

Your choice of punctuation depends in part on whether you prefer open punctuation (use it where it will help the reader) or close punctuation (punctuate wherever the rules permit). The purpose of punctuation, however, is not to be "correct" but to help your readers. Ideally they should be able to hear your voice in the prose. You know better than anybody where you would pause, where you would come to a full stop.

A simple way to punctuate, then, provided you are a reasonably good reader, is to read the prose aloud and listen to your voice. You naturally pause where a comma is needed; you come to a full stop for a period or a semicolon (which has the effect of a period but relates the ideas more closely). You hear parenthetical expressions that call for parentheses or dashes. Dashes must be used carefully, though; in some contexts they sound childishly breathless. Exclamation marks have a similar effect and are rarely appropriate for technical or scientific writing.

One thing you do not hear as you read is the hyphen, and hyphens are important marks of punctuation. They affect meaning. Observe how hyphens clarify the meaning (or change it) in the following phrases:

(a) heavy oil field equipment = heavy oil? or heavy equipment?

heavy-oil field equipment = field equipment for a heavy-oil operation

heavy oil-field equipment = heavy equipment for an oil-field

(b) chip carrying capacity = capacity of something to carry chips? or capacity of chips to carry something?

chip-carrying capacity = capacity of something to carry chips

chip's carrying capacity = capacity of a chip to carry something

(c) demodulator filter terminating system = ?

(d) crude oil carbon dioxide system foaming = ?

See the entry on *hyphens* in the "Dictionary of Common Problems" for some rules that help.

11. When Does One Use Numerals?

And when must a number be written out? See the rule and its many exceptions in the "Dictionary of Common Problems" under "Numerals."

12. Mr., Mrs., or Ms.? Dear Sir, Madam?

(a) In the address, you can omit the title if you don't know whether it's Mr., Ms., Mrs., or Miss: J.L. Johnson; B. Richardson.

(b) In the salutation you can sometimes use professional titles: <u>Dear</u> <u>Board</u> <u>Member</u>: <u>Dear</u> <u>Professor</u>:

(c) Or you can omit the salutation entirely. (When you get right down to it, you aren't writing to a "dear" anybody.) You may have noticed that many contemporary letters are written this way. If you omit the salutation, you can also omit the complimentary closing. Terms like "Very truly yours" and "Respectfully yours" are stiff traditional endings handed down from the days when one literally belonged to a noble or royal personage. We don't really mean "I am yours."

(d) If you don't know whether your reader is a Mrs. or a Miss, no problem; a woman's marital status is as irrelevant as a man's. Ms. applies to all women. (See "Sexism in On-the-Job Writing," p. 246.)

Dictionary of Common Problems

We suggest that you scan the following entries at least once. Some of them may flag problems you are unaware of having. For instance, the writer who knows no better than to say "John and myself" would never think to look for that condemned practice under "myself"; nor would a person who pronounces *asterisk* "asterix" deliberately go looking for advice on it.

adverse, averse   Since *averse* means reluctant or disinclined, it refers only to people (or animals). *Adverse* (hostile or unfavorable) refers to conditions.

affect, effect	*Affect* is always a verb meaning "to produce an effect upon; to alter or change." *Effect* is either a noun (that which is produced or brought about) or a verb (to bring about, cause, or implement). "I may effect [produce] a change in my behavior; the effect [change brought about] will affect [alter or change] my life."
agree to, agree with	Agree *to* a proposal; agree *with* a person. "In agreement *with*" in all cases.
agreeable to	Not "agreeable *with* me."
agreement between subject and verb	See *number*.
all right	Correct. *Alright* is a non-word.
amount, number	Use *amount* for an aggregate that does not have countable units. Use *number* when the mass or group can be separated into countable units. ("The amount of sand" but "The number of sandstones.") See *fewer, less.*
and/or	Use it carefully, as it can create ambiguity. Usually *and/or* boils down to *or.* See *virgule.*
ante-, anti- (prefixes)	*Ante-* means "before"; *anti-* means "against" or "opposed." Both are usually solid with stem: antedate; antibody; antifriction. But they are hyphenated when joined to a capitalized word (ante-Victorian; anti-Roosevelt) and when *anti-* is joined to a word beginning with *i (anti-inflation).*
a number of, the number of	*A number of* is always plural; *the number of* is singular. ("The number of people responding was disappointing." "A number of people were not willing to respond.")

as, because   Be careful about using these as syn-
                                   onyms, for they can be ambiguous.
                                   ("We began a new project as the first
                                   one was nearing completion." Mean-
                                   ing: during that time? or because the
                                   work was nearing completion?)

as far as, . . . is   "As far as" is incomplete, ungrammat-
concerned   ical, and illogical without "is con-
                                   cerned."

as, like   Despite the rearguard action by most
                                   grammarians, *like* will eventually be ac-
                                   cepted as a conjunction in formal usage,
                                   and Shakespeare's play could become
                                   Like You Like It. Meanwhile, the care-
                                   ful writer will stick with *as*.

as well as,   These terms do not make a singular
along with,   subject plural. ("The author, along with
together with   many readers, was pleased with the re-
                                   port.")

at present, presently   *At present* means "now" or "at this
                                   time." *Presently* means "before long"
                                   or "soon." (Some writers imagine that
                                   presently is a fancier word for *now*.)

co-   Nearly always solid with stem in mod-
                                   ern style: cooperation; coordinate; co-
                                   exist. Exceptions: co-opt, co-ops, co-
                                   worker. See the dictionary if in doubt.

colon (uses)   Colons are used to introduce a list, se-
                                   ries, quotation, or example. They may
                                   also be used to separate a clause from
                                   another clause that restates the idea in
                                   different words.

comma (uses)   Use commas to:
                                   (1) Separate ideas of equal rank
                                       (clauses joined by *and, but, or,
                                       nor, for*) unless the clauses are very
                                       short.

(2) Set off introductory phrases, as a rule. Example: "Using her best judgment, she selected one of the methods." To check the punctuation, read the sentence aloud.

(3) Set off introductory dependent clauses: "Before we decide on a plan, we must study all options offered."

(4) Separate elements in a series, including the last element before the *and*. ("The colors were brown, yellowish brown, and yellow." "The combinations chosen were yellow and brown, gray and yellow, and yellow and grayish-green.") Note: if the elements are complex within themselves, use semicolons instead of commas between them.

(5) Set off nonrestrictive dependent clauses (nonessential, added information). "Students may fly all aircraft, which have recently been repaired and serviced." But if the clause is restrictive (i.e., it contains information essential to the meaning of the main idea), then omit the comma and use *that* instead of *which*: "Students may fly all aircraft that have been repaired and serviced."

compare to, compare with

Compare with for similar objects or qualities. ("Compare dead happiness with living woe.") *Compare to* for dissimilar things. ("Compare man to machine.")

compose, comprise

Compose = to make up. ("Fifty states and several territories compose the USA.") *Comprise* = to include or embrace. ("Canada comprises 10 provinces and 2 territories.")

criteria, criterion   *Criteria* is plural: "The criteria are as follows." *Criterion* is singular: "A good criterion is available."

dangling constructions   See p. 183.

data is, data are   See p. 184.

de-   Usually solid with stem unless joined to word beginning with *e*: demagnetizer, degasser, desurger; but de-energize, de-emphasize.

different from,   *Different from* will never be criticized.
different than   *Different than* will be criticized by many careful writers, though its use grows.

disinterested,   *Disinterested* = impartial, free of bias.
uninterested   *Uninterested* = not interested, indifferent.

due to, because of,   Preferred use of *due*, which is an adjec-
owing to   tive, is with a noun. ("The failure was due to," not "It failed due to.") Use *owing to, because of,* or *on account of* with verbs. ("The material burns because of," not "burns due to.") Note: some usage authorities no longer condemn using *due to* with a verb, although they wouldn't consider doing it themselves. If you are uncertain about a particular use of *due to*, substitute *caused by*. If it makes sense, then *due to* is fine. And don't start a sentence with *due to*.

effect, affect   See *affect, effect*.

equally as   Redundant (Fowler goes so far as to call it "an illiterate tautology"). Drop either *as* or *equally*.

farther, further   Use *farther* for physical distance. Use *further* for everything else.

fewer, less   Use *fewer* if you speak of countable

units; use *less* for amounts that cannot be divided. (<u>Fewer</u> people, <u>less</u> material; <u>fewer</u> calories, <u>less</u> fiber.)

firstly, secondly

Why not *first* and *second*? What does the *-ly* do for us?

hopefully (adverb)

Its use to mean "It is hoped that" or "We hope" is not advisable for technical, scientific, and business writing. ("Hopefully, you will not follow this bad example.") From the *American Heritage Dictionary*: "<u>Hopefully</u> . . . is still not accepted by a substantial number of authorities on grammar and usage . . . ." The word "still" in that sentence conveys the reality that the growing use of the word as an adverb for the entire sentence will eventually make the usage acceptable. This has already happened with the adverb *fortunately*.

hyphens

Some general rules:
(1) Hyphenate compound (unit) adjectives. ("I want an air-conditioned room.")
(2) Hyphenate compound verbs. ("We must air-condition the computer room"), but do not hyphenate verbs followed by a preposition ("They cut-off the power" is wrong).
(3) Hyphenate nouns of three or more words (right-of-way); nouns composed of verb plus *-er* plus adverb (runner-up); compound units of measurement (acre-foot); letter-noun combinations (I-beam).
(4) Hyphenate a noun when the first word ends in *'s* (*bull's-eye*).
(5) To hyphenate compound nouns, consult the dictionary.
(See discussion of hyphens on p. 189.)

if, whether,

If and *whether* are interchangeable words. But *if* should not be followed by *or not* (WRONG: "I don't know if this is a good solution to the problem or not.") *Whether* is generally used without the *or not* unless alternatives must be stressed: "We will pursue the new research whether they support us or not."

imply, infer

The writer implies, the reader infers.

inter-, intra-

Usually solid with stem, no hyphen: interglacial, interdepartmental, intradepartmental.

invention of new words

Do not invent words when precise ones already exist: e.g., *keyboarding* for *typing; gifting* for *giving; vugular* for *vuggy*.

its, it's

Its is possessive; *it's* is a contraction for *it is*.

lend, loan

Careful writers prefer *lend* for the verb and *loan* for the noun. ("I'll lend you money, but you must repay the loan in six months.")

lie, lay, lain

Lie is an intransitive verb; that is, it does not take an object: "The building site lies [present tense] in the valley; it lay [past] in the valley; it has lain [present participle] in the valley for centuries." *Lay* is a transitive verb; it does take an object. ("We will lay the foundation today; we laid it yesterday; we have laid six foundations this week.") If you say, "I laid there all day," you imply that you are a hen.

like

"I feel like it will work" is a solecism for "I feel that it will work." (See also *as, like*.)

media is, media are   *Media* is plural: use the plural verb *are.*
                          Medium is singular: use the singular
                          verb *is.*

modifiers   (1) Place adjectives and prepositional
                          phrases as near the words they
                          modify as possible. *Both* and *only*
                          are commonly misplaced:
                          This: "We plan only to start it."
                          Not this: "We only plan to start
                          it."
                          This: "It covers a wide range of
                          both age and geography."
                          Not this: "It both covers a wide
                          range of age and geography."
                          (See p. 163: Link Ideas to Show
                          Equal Rank.)
                          (2) Avoid too many prepositional
                          phrases in a string:
                          This: "The purpose of the study is
                          to see if the group in this area can
                          be zoned by microfossils."
                          Not this: "The purpose (of the
                          study) is the evaluation (of the pos-
                          sibility) (of zonation) (of the group)
                          (in this area) (by means) (of micro-
                          fossils)."
                          (3) Break up noun trains (see p. 151).

myself, me, I   The use of *myself* instead of *I* or *me* is
yourself   the prissy (and unsuccessful) way some
                          writers try to avoid personal pronouns:
                          "Jones and myself [should be *I*] at-
                          tended the meeting." "The committee
                          appointed Jones and myself [should be
                          me] for the field study." You would not
                          say "myself attended the meeting" or
                          "the committee appointed myself."

non-   Usually solid with stem. But hyphenate
                          when *non-* is joined with a capitalized
                          word.

none is, none are   Both are correct. See usage note in the *American Heritage Dictionary*[1].

not only, but also   Should be followed by constructions that are parallel in form; see p. 163 for a discussion of parallelism.

number (agreement between subject and verb)

(1) The verb should agree with the subject, not with intervening modifying nouns: "A <u>set</u> of tools <u>was</u> ordered."

(2) If a subject is composed of one singular and one plural noun, make the verb agree with the nearest noun: "Neither the chairman nor the <u>members</u> <u>were</u> satisfied."

(3) Singular nouns having plural modifiers that indicate additions to the subject (*together with, along with*, and *as well as*) take singular verbs. ("The <u>chairman</u>, as well as the members of his committee, <u>makes</u> this recommendation.")

(4) *A number of* is usually plural; *the number of* is singular.

(5) Some nouns are either singular or plural, depending on their use. *Data are* means the individual pieces of data; *data is* makes them a body of data (a collective noun). See p. 184.

numerals (use in text)   The general rule: Write out figures below 10; express figures 10 and above in numerals. "There were only nine men to run 85 machines." Exceptions:

(1) Don't begin sentences with numerals. If the number is big (1,485,690), recast the sentence to remove it from the beginning.

(2) Several numbers cited in a short space can be expressed consistently as numerals: "Of the 49 partici-

pants, 40 passed the test, 2 fell below 60 percent, and 7 dropped out.''

(3) If a figure below 10 is closely associated with one above 10 in meaning and position, both may be written as numerals: "The values were 92 and 2."

(4) Occasional round numbers may be spelled out: "This occurs every hundred years or so."

(5) Use numerals with abbreviations for units of measurement: "6-km run"; "⅝-in. tubing."

on, upon　　　Most editors prefer the simpler *depending on; depending upon* is also correct.

oriented, orientated　　　They have the same meaning, so why use the bigger word?

out of, of　　　As in "this piece is made out of metal"; drop the *out*.

parallelism　　　See p. 163.

past history　　　Redundant; history is always past.

percent, percentage　　　Partially synonymous. Use *percent* with a number to indicate a specific fraction ("20 percent"); use *percentage* to indicate an indefinite fraction ("the largest percentage").

personal pronouns　　　See *myself*, above; check for False Objectivity (p. 148).

phenomena, phenomenon　　　Keep the distinction: *phenomena* is plural; *phenomenon* is singular. (See *criteria*.)

possessives　　　The apostrophe follows the *s* for plural possessives: "two days' time"; "five years' work." Exception: When the word is made plural by internal change in form (men, women), the apostrophe

goes before the *s*: men's, women's. Use
's for singular possessives: "a week's
job; a month's time."
Note: When possessives are made unit
modifiers, they are hyphenated and the
apostrophe is omitted: "A one-week
course"; "a six-month study."

presently   See *at present*.

proved, proven   Both *have proved* and *have proven* are
                                 acceptable. *Have proved* is the more
                                 modern usage.

real, really   *Real* is an adjective and cannot be used
                                 in formal writing to modify an adverb.
                                 Write "He really works hard" or "He
                                 works really hard," not "He works real
                                 hard."

reason is because   WRONG. "The reason we dropped the
                                 project is that [not because] our analyst
                                 resigned."

semicolon   Some general rules:
                                 (1) Separates main clauses without a
                                     conjunction (has the effect of a pe-
                                     riod but relates the ideas more
                                     closely).
                                 (2) Separates elements in a series when
                                     each element is very complex or
                                     when each has commas internally.
                                     ("The report contained a descrip-
                                     tion of the machine, beginning with
                                     its components and building up to
                                     the whole; a series of functional
                                     diagrams, each designed for the
                                     new operator's understanding; and a
                                     section describing applications.")

shall, will; should,   See p. 187.
would

since, as, because   *Because* is the strongest word to express
                                 cause. *Since* is a good synonym, pro-

vided it does not suggest time in context. (Archeologists, geologists, and historians must use it very carefully.) See also *as, because*.

slash, stroke See *virgule*.

split infinitives See p. 182.

that, which Many technical writers seem to feel that *which* is more formal than *that*. In fact, the two pronouns are used for different purposes in introducing subordinate clauses.

(1) Use a comma followed by *which* to introduce information that is not essential to the meaning of the main idea: "You may use any of the computers, <u>which</u> have at least 256K RAM."

(2) Use *that* without a comma to introduce information necessary to the meaning of the main idea: "You may borrow any of the computers <u>that</u> have at least 256K memory." (Don't borrow the ones with insufficient memory.)

they, their, them (with singular verbs) See "Sexism in On-the-Job Writing," p. 247.

try and, try to Make it *try to. Try and do it* means two operations: try it, and then do it.

under way Two words.

virgule (slash) The slash (*stroke* in British usage), as used increasingly by people in too much of a hurry to decide exactly what they mean, creates ambiguity. For instance: *London/Paris* means London **to** Paris? London **and** Paris? London **or** Paris? *Runaways/orphans* could mean "runaways, some of whom are orphans";

"runaways **or** orphans"; "runaways **and** orphans"; or even, conceivably, "orphans, some of whom are also runaways." Don't leave the reader guessing.

whether, whether or not

See *if, whether*

who, which, whose

Refer to people as *who*, not *which*. ("Our president, who wrote these policies, has retired.") Reserve *which* for inanimate objects. Use *whose* for either.

who, whom

Use *who* as the subject of a clause; use *whom* as the object of a verb, preposition, infinitive, gerund, or participle. See p. 184.

will (after if)

Like *would have*, this is a common problem for people using English as a second language. Write "If tomorrow is rainy, the test will be cancelled," not "If tomorrow will be rainy, the test will be cancelled."

-wise

Terms such as *efficiency-wise* and *economy-wise* are slangy fads that the *wise* writer declines to indulge in. Theodore Bernstein calls attention to a cartoon in *Punch* showing owl parents discussing baby owl; one says to the other, "How's he shaping up, wisewise?" (Bernstein commented, "If ridicule could kill, that cartoon should have had the '-wise' fad lying lifeless at our feet.")

would have, had (after if)

"If we had [not would have] gone to the airport . . . ."

*   *   *   *

This dictionary is not intended to cover mispronounced words, although they can be as distracting in a spoken presentation as misspelled words in a document. Two words merit mention here because they are so frequently mispronounced by otherwise literate speakers:

asterisk   The word is pronounced *as-ter-isk*, as in the word *risk*. "Asterix" sounds illiterate.

nuclear   The word is pronounced *nu-cle-er*, never *nu-kew-ler*.

Part IV

International Scientific and
Business Writing

A GLOBAL LINGO

Approximate number of people using English as a first language	345 million
Approximate number of people using English as a second language	400 million
Percent of international telephone conversations in English	85
Percent of technical papers (world-wide) published first in English	80
Percent of world's secondary school students studying English	77
Percent of information stored in English in the world's computers	75

Figures quoted are from *U.S. News & World Report*, Feb. 18, 1985, pp. 49-50; *Harper's Magazine*, April 1985, p. 11; and Richard Reeves in the *Houston Post*, Mar. 9, 1983, p. 2B.

Part IV
International Scientific and
Business Writing

Under the twin effects of electronic communications and jet travel, our world has shrunk to a global village (in Marshall McLuhan's term), and it will shrink further. Yet reaching an understanding with faraway readers can still be a problem. Our approach to the problem, in this part of the book, goes right back to Step 1 of the method—know your audience. This is the essential starting point for effective on-the-job writing to people in other lands.

DIFFERENT CULTURES, DIFFERENT CUSTOMS

In spite of the worldwide use of English (see opposite page), written communication in English does present special problems when you are involved in science or business overseas. This is as true for North Americans writing to foreigners as it is for other people writing to North Americans.

Words, which are supposed to enable us to get through to others, can sometimes cause misunderstandings because of the way they are used. The manner in which we phrase a message can be of critical importance, especially in dealing with other cultures. We may be careful to state precisely what we mean, but even a reader skilled in English can receive a different message if his cultural traditions have fostered different expectations with different implications from a given sequence of words. (In languages related to English, there is the additional problem of "false friends" [p. 233].) At best we may unknowingly frustrate, and at worst we may anger, a foreign reader.

In this sort of situation, writers often wonder whose expectations have priority. Whose standards of usage should they observe? What guidelines can they follow? Overseas clients will sometimes ask us, "Why should I be compelled to use Americanisms? Why do you expect me to write the way an American would?"

The answer: nobody is compelled to follow another country's usage. But if you are, for example, an American writing to the head of your company in London, it would be considerate to indicate measurements in the units commonly used in the British sector of your industry and to express dates in British style. If you are an Australian writing to an American in the head office in San Francisco, you can help him feel comfortable with your message by employing American usage, without feeling anxious about any lapses you might have. On the other hand, if you are an Australian writing to another Australian—even within the subsidiary of an American company—you should of course employ standard Australian usage.

Whether you are a North American writing to people somewhere else, or a professional using English as a second language to write to North Americans, the problems from either side can be handled by giving special thought to Step 1 (page 29): Analyze Your Readers. Although every case must be handled individually, in this section we review some of the most common situations and suggest ways to handle them.

American vs British English

An especially easy case of writing to foreign readers is that in which Americans are writing to readers whose first language is British English, and vice versa. By "British usage" we mean that espoused by H.W. Fowler[6] and used today in the U.K., Australia, and most other Commonwealth countries.

A Commonwealth exception is Canada, where variations from American usage are few and slight:

- Canadians have switched over to SI (Systeme International) units of measurement.
- Some Canadians favor British spelling (favour, colour), except in the case of verbs ending in *-ize*. British and Australian printers commonly use *-ise,* as in *industrialise,* although Fowler, the Oxford University Press, and the Cambridge University Press are all against it.
- Canadians show "March 10, 1989" as 10/3/89, the way Europeans do, or, with increasing frequency, as 89/3/10.

With these exceptions, Canadian usage in scientific and business writing is pretty well indistinguishable from that south of the border.

On the other hand, between British and American usage there

Which Metric System?

If you have analyzed your readers, their backgrounds and needs, you should already know whether it will be most efficient to use British units, "old" metric units, the new standard SI (Systeme International) units, certain traditional local units, or a mixture of several systems.

Usage varies from industry to industry and from country to country with little internal consistency. In the U.S. health industry, for example, perhaps five out of every six measurements are expressed in metric (in Canada, it's metric 100%). The only important remaining use of British units in the field of American medicine is in communicating with patients; height, weight and temperature are usually expressed in the old units.

In the "oil patch," as another example, some countries otherwise converted to metric units still use *barrels, feet,* and *pounds per square inch.* In others, like Canada, the petroleum industry has by government fiat converted to SI units: *meters, newtons,* and *pascals.* In still other oil-producing countries, units may be an interesting mix of *meters* (for well depth), *barrels* (for volume), and pre-SI metric units like *centipoises* (for viscosity).

As a courtesy to a varied readership, you can show a value in the most commonly used unit first, followed in parentheses by its equivalent in the second-most-familiar unit. If you are obliged to use an unfamiliar system, there are plenty of publications you can turn to for help. A good example is *Guidance for Using the Metric System SI Version* by Valerie Antoine (1975). Many professional societies publish reference manuals providing help with the units used in their particular areas of science or industry.

are enough differences to fill a good-sized glossary. Two terms that are still current in British and Australian usage deserve mention. The conjunction *whilst* and the business-letter closing *Yours faithfully* (or *Faithfully yours*) have not been used in North America for many, many decades. In consequence, they sound pretty antiquated to American (and Canadian) ears.

Watch out also for *billion* and *trillion,* which in Britain and Germany mean 10^{12} and 10^{18}, respectively, while in France and the U.S. they mean 10^9 and 10^{12}, respectively.

Transatlantic differences in spelling are, in reality, inconsequential. If you prefer to spell it *colour,* by all means do so. If you prefer *color,* feel free to use it. Nobody will misunderstand you. If you are writing for another's signature, spell it his way.

Muslim Cultures

In writing to people of any Muslim culture, the main points to keep in mind are, first, their high level of courtesy in formal documents, and second, the inseparability of religion and daily business. In Arabic correspondence, devout Muslims often invoke the blessings of God in both the opening and closing of a business letter. Even if you could do this sincerely, it would not be expected unless you were a Muslim yourself. To establish a rapport with your reader, depend on respectfulness and friendliness.

In lieu of the traditional opening, we can include a polite sentence or two in a brief opening paragraph to establish a pleasant relationship. To jump in and start discussing business before this courtesy has been extended might appear impolite.

East Asian Cultures

In writing, as in conversation, people of eastern and southeast Asia (who do not care to be referred to as "Orientals") consider indirectness a fundamental aspect of courtesy. One expert, a former U.S. ambassador to Japan, commented:

> Certainly the Japanese, with their suspicion of verbal skills, their confidence in nonverbal understanding, their desire for consensus decisions, and their eagerness to avoid personal confrontation, do a great deal more beating around the verbal bush than we do, and usually try to avoid the 'frankly speaking' approach so dear to Americans. They prefer in their writing as well as their talk a loose structure of argument, rather than careful logical reasoning, and suggestion or illustration, rather than sharp clear statements.*

---

*E. Reischauer, *The Japanese* (Cambridge: Harvard University Press, 1977), p. 386.

A Japanese engineer in one of our classes, who had buried his main recommendations deep within a report, defended himself this way:

> "Even though I have been working in the United States for several years, it is still difficult for me to put my recommendations prominently at the beginning of a report or memo. If I did that in Japan, it could provoke a confrontation. Suppose senior management did not agree with my new proposals? I would be forced to withdraw my report, or at least that portion of it, and I would lose face. If I were rash enough to insist on the recommendations, my boss would lose face. So, to avoid confrontation it is best, in my country, to imply new recommendations in a subtle way."

We do not recommend so deep a bow to foreign custom. Indeed, some East Asians (those of Hong Kong, for example) have largely overcome the unwillingness to risk disagreement. But if you are writing a business letter to someone in the Far East, compromise. Aim for an appropriate level of courtesy in the opening paragraph, and then get to the point. You don't need to tie yourself in knots trying to be indirect. Tact requires a little extra planning in organization, a little courtesy, and respect for the sensibilities of the reader.

Latin American Cultures

Since Latin Americans are our closest non-English-speaking neighbors, communicating with them in writing ought to be comparatively easy. Still, it helps to keep a few cultural differences in mind.

Business writing in Latin America differs from that in North America in four principal ways whose implications are discussed below.

- Many Latin Americans take a less-hurried approach to business writing.
- Traditional courtesy is still very much alive in Latin American correspondence.
- You might suspect that personal relationships, whether casual or very close, would set the tone for a letter; but while this is

true of personal correspondence, it is not true of business writing.

■ Latin Americans are keenly aware of, and very sensitive to, differences in job level between writer and reader.

Not being in a rush means that there is time for traditional courtesy, expressed both in the opening and closing of a letter. The words employed may strike a North American as flowery and perhaps "unbusinesslike," but they are part of the standard ritual, exactly like our salutation "Dear . . ." in letters to total strangers and closings like "Very truly yours." However, though Latin Americans may expect the full treatment from their compatriots, they do not necessarily expect it from foreigners. When writing to Latin Americans in English, you can take time for a "courtesy" sentence, if it's appropriate, and then launch into business.

While personal relationships (even casual acquaintanceships) can be very important in Latin America, they scarcely affect the tone of business writing, which is invariably formal and business-like. In business writing to people who happen to be friends or acquaintances, Latin Americans rarely permit themselves the informal tone that we could adopt under comparable circumstances. Informality is reserved for personal letters.

Formal courtesy is also reflected in correspondence between people of different job status. North Americans do well to keep this in mind when writing to government officials of any rank. Without being flowery, the tone of the correspondence should be formal, implying a degree of respect that we would find exaggerated in addressing our own civil servants, however exalted they might be. The reason for this is easy to appreciate: in dealing with foreigners, Latin Americans may feel especially sensitive about their situation as official representatives of their government or institution, thus quick to detect any lack of respect on the part of a foreigner.

WRITING TO PEOPLE WHO USE
ENGLISH AS A SECOND LANGUAGE

Although certain niceties apply to correspondence with all ESL readers,* each culture has its own literary mannerisms. The general guidelines are:

---

*ESL readers: readers using English as a Second Language.

- Prefer simple sentences and easy words.
- Be sensitive to the correspondence etiquette of your reader's country.

First Priority: Simplicity

While it is not always possible to know your audience well, in communicating with foreigners it certainly helps to have some idea of their command of English. With that you can tailor your letter, memo, report, or speech to the appropriate level. It is a truism that educated foreigners quite often speak and write a more grammatical English than the average North American, and their vocabularies may exceed our own.

There's an old story, probably apocryphal, about V. Wellington Koo, Chinese ambassador to Britain in the 1920s. One night at a diplomatic function in London, he was seated next to an English lady for dinner. The lady spoke no Eastern languages but felt she could not ignore her neighbor entirely, so she smiled at Mr. Koo and asked, "Likee soupee?" Ambassador Koo limited himself to a smile and a nod. Relieved, the lady turned to converse with the gentleman on her other side.

Later in the evening, the ambassador was called on to address the guests. He spoke brilliantly without notes for about 15 minutes, while the lady squirmed. When he sat down, he reportedly turned to her with a smile: "Likee speechee?"

A little investigation (Step 1) can save us this sort of embarrassment. In general, you can help foreign readers by keeping your sentences to a reasonable length and by preferring common words. Easy words are usually—but not always—short words. In particular, avoid slang, puns, metaphors, and colloquial sayings that may leave your readers feeling flummoxed (as they could be by that last word).

A good way to achieve simplicity is to imagine that you are in a face-to-face conversation with your foreign reader. This will help you keep the reader's language problems in mind.

Courtesy in Business Letters

In a business letter, the North American (and to a similar extent, the British) tendency is to dispense with the formalities and get down to brass tacks. We did away with ceremonious greetings so long ago that the following opening, translated from a modern business letter in Spanish (chosen as a typical example), strikes us as flowery to the point of travesty:

> Very Esteemed Sir:
>
> By means of the present (letter) we are very pleased to address ourselves to your worthy institution, to greet you very respectfully and to ascertain if it might be possible to obtain some information related to . . . .

In English translation, the letter closed like this:

> With nothing further to report at the present time, we express our grateful thanks for your appreciated cooperation in this very important matter, and remain,
>
> Your eternal and faithful servants,

North Americans writing in English should not try to reproduce this old-fashioned courtesy. It is well, however, to consider how our typical breezy opening may sound to someone whose culture still calls for a more formal approach. We may unintentionally come across like this:

> Hey, you!
>
> How about helping us with a project in . . . .

Quite possibly, what we see as a laudable effort to be open and frank may be perceived as loud, even pushy. What we view as time-saving may be seen as tactless. In many countries it is much more important to save face than to save time.

This calls for an effort to exercise tact. Write (or edit) so as to cast sentences in terms that take the harshness out of criticism. Never let a statement make the reader feel foolish or stupid. Edit the aggressive tone out of differences of opinion; recognize the reasonableness of the reader's viewpoint before refuting it in gentle but plain language.

This kind of writing is a compromise between the no-nonsense, get-to-the-point directness we recommend in this book and the elaborate courtesies foreign readers might prefer. In teaching ESL writers addressing North American readers, we ask for a similar compromise.

In both cases, the object is to write as plainly and directly as possible without offending the culture-based expectations of the reader. In either case, it is not only courtesy that is involved, but also self-interest. We are talking about understanding, respect, willing cooperation, and good business relations.

FOR WRITERS USING ENGLISH AS A SECOND LANGUAGE

If you have trouble expressing yourself in English, you are not alone. Of the nearly half-billion people who use English as a second language, a great many are frustrated by its inconsistencies (to say nothing of the native English speakers who have difficulty expressing themselves in writing). Most of this section was developed from our international workshops in technical writing.

You Have a Built-in Advantage

Strange as it may seem, using English as a second language can actually give you an advantage over writers whose first language is English. The reason: you are forced to write simply. And in modern business and scientific writing in English, simplicity is preferred.

If you do not have a vast vocabulary and a command of complicated sentence structures, you will not be able to write gobbledygook. The dictionary defines *gobbledygook* as "unclear, often verbose, bureaucratic jargon." The word is a combination of *gobble* (to sound like a turkey) and *gook,* which may have been derived from the Scottish *gowk,* meaning a simpleton, a silly or stupid person.

You need not worry, then, if your language is simple. Readability studies show that most readers, including college graduates, like it that way.

On the other hand, it must be said that even simple sentences can be garbled so as to become unintelligible, and nobody likes that. Sentences must follow recognizable patterns, and words must express specific meanings that say the same things to everyone. Preferably both words and sentences will conform to established educated usage in writing.

The following sections are directed to ESL writers who prepare reports, letters, memorandums, and professional papers for English-speaking readers. As might be expected, some attention is paid to our North American style, beginning with the following section.

You Can Get to the Point

Along with economy of style, you can achieve economy of organization. In writing to North Americans, you do not have to practice what we call "beating around the bush"; that is, you do not need to delay your NEWS.

ESL writers are sometimes concerned about being too blunt in putting their NEWS "right up front." They are not aware of the important difference between being *brisk* (quick) and *brusk* (discourteously abrupt). Briskness in this sense is a virtue of direct organization. Bruskness, on the other hand, is a fault in the choice of words, largely independent of organization.

Brisk writing saves time and effort (as well as money) for many busy people in the working world. Within an organization, they include the writers themselves, supervisors, managers, top executives, other addressees, and recipients of copies. Adding readers outside the organization, the savings may accrue to dozens of readers.

Within the last few decades, the business and scientific community has expanded greatly to include growing numbers of peoples of different cultures and languages. Communication between these peoples presents increasing danger of misunderstandings and bad relations. Thus it is understandable that plain language and efficient organization are considered courteous by English-speaking readers everywhere. They appreciate your getting to the point.

If your cultural background makes this kind of writing very difficult or distasteful in your view, try a compromise. Write a "polite" opening paragraph (but keep it brief), and then get briskly to the point for businesslike efficiency and assured understanding.

Avoid Poor Models

ESL writers tend to model their writing in English after that of a senior colleague or a superior. This is an understandable approach but a risky one; the model's English may be no better than the writer's.

Perhaps your boss is also an ESL; perhaps his command of English seems impressive to you; and perhaps you feel, consciously or unconsciously, that imitation is the sincerest form of flattery. But if your English is not good enough to judge the boss's competence as a writer, you may simply end up copying his mistakes and picking up his bad habits, in a fine example of "the blind leading the blind."

Copying the writing of a North American colleague or boss is not always a good idea, either. It is sometimes difficult for foreigners to realize that many North American college graduates struggle with deficiencies in their own writing of which they are keenly aware. They would like to write better English, but they are stuck with their limited best. They would react with amusement or dismay if they knew someone was using them as a worthy model.

This does not mean that you, as an ESL writer, must go it alone. It does mean that you should learn from qualified books and teachers rather than from people who make no claim to expertise in English.

Common ESL Problems

The major problems of people writing in English as a second language tend to fall into the categories reviewed below. If we have not covered your special problem in this section, consult your English instructor or textbook.

As you have heard before now, the verb is the most important word in an English sentence. It controls the clarity, directness, pace, and style of one's writing. When you learn to handle verbs efficiently, your writing difficulties are well on the way to being solved. Two of the following sections deal with verb problems; others are treated in the "Dictionary of Common Problems" (p.190).

Idiomatic expressions

Idioms, by definition, are peculiar. They are expressions that follow no consistent logic and must therefore be learned without reference to helpful rules.

For example, Table 5 shows how a preposition tacked onto a verb can totally change the meaning of that verb. A great many of these combinations originated as slang; some of them are still slangy. Although the table includes more than 80 common examples of this peculiarity of English, it would take an entire book to cover them all.

To use the table, combine any verb from the lefthand column with any preposition in the other column headings and read across to the various meanings. If you cannot find the combination you seek in this table, try the dictionary or the index section of *Roget's Thesaurus*[2].

You will observe that the prepositions in Table V follow the verbs. These combinations should not be confused with those in

TABLE 5.
Examples of Different Meanings for Verbs When Followed by a Preposition

VERB	PREPOSITION					
	about	*around (round)*	*away*	*back*	*down*	*in*
break			desert, escape; become detached	revert, regress	get out of order; classify; weep; collapse	enter forcibly; interrupt
bring	cause; accomplish	revive		fetch, return; restore, revive	fell; raze; humiliate	introduce
come	happen; change course (a ship)	agree, acquiesce; recuperate; visit	leave, depart	return; recur; recover; answer	descend	arrive, enter
get	move, circulate	evade, outwit; be publicized; travel	escape; depart	recover; (+ *at*) = revenge oneself	descend; lower; dismount	enter; arrive; participate
go	wander, circulate	rotate; circle; circulate; suffice	depart; disappear; recede	return; retreat; revert	descend; decline; diminish; fail	(enter)
put	turn, change course (a ship)	distribute; circulate (news or rumor)	store, secrete; kill (slang); eat (slang)	return; restore; delay	record; suppress; disparage	contribute
take		distribute	remove, subtract; lessen, detract	recant, apologize; recover; return	record; raze, bring down; disassemble	accept, receive; include; hear; deceive (sl.)
turn	reverse, invert	(rotate)	deflect; dismiss; reject	repulse; retreat	reject	deliver; go to bed

sl = slang

PREPOSITION					
off	*on*	*out*	*over*	*through*	*up*
discontinue; interrupt		erupt; escape; begin		penetrate; make good; succeed	disintegrate part company analyze
succeed (with) accomplish; carry out	induce, incur	elicit; indicate; issue, print			propose; train, nurture; vomit
occur; become detached	appear; approach	eventuate; bloom (a flower); debut	pass; pass over	penetrate; be understood; succeed	ascend; increase; occur
dismount; depart; go free; begin	progress; mount; age;	leave, escape; extract; issue, print	recover (from); move; be understood	penetrate; complete; be heard	arise; prepare; raise
explode; depart	begin; continue, persist	(leave); be extinguished	examine; rehearse; study; succeed	experience; search; spend ($);	(ascend); increase
postpone, avoid; annoy	wear, assume cover; apply	exude; extinguish; annoy	succeed; deceive	execute; channel	erect; contribute
					(+ with) = tolerate
detach, remove; ascend; imitate leave } sl	assume; undertake; fight with; employ	remove, exclude; extract; escort	acquire, assume command; appropriate		discuss, deal with; espouse; raise; engage in
extinguish; stop (a machine); discourage	start (a machine); excite (sl); take drugs (sl)	evict; switch off; attend; eventuate	deliver; transfer; rotate		appear, arrive; find

which the preposition precedes the verb. For instance, *overcome* is unrelated in meaning to *come over*.

English also makes one-word **nouns** out of verb-plus-preposition combinations. For example, from the verb *to break down* we have the noun *breakdown*, meaning a collapse. To *put down* becomes a *putdown*; *to take over* becomes a *takeover*.

Sometimes nouns develop with the preposition preceding the verb: from *put through* (verb) we have a *throughput* (noun). Some nouns duplicate the verb form: *to overload* (verb) becomes an *overload* (noun); *to overthrow* becomes an *overthrow* (noun).

Note: The small pocket dictionaries such as Arabic-English, French-English, etc., are not sufficiently inclusive for the serious writer. An unabridged dictionary (and a thesauraus) belongs on every writer's desk.

Sentence structure: the skeleton

English sentence structure is highly varied; it can present problems even for people whose native language is English. If you have difficulties with sentence structure, learn to "pick the skeletons" of troublesome sentences.

The skeleton is the basic part of the English sentence: the subject, verb, and complement, if any, of a clause (idea). Within the clause, every element other than the skeleton is subordinate; that is, it serves only to describe, limit, or define a word in the skeleton. In informative writing (business, technical, or scientific), it is especially important that the skeleton carry the main part of the sentence's message.

The technique for picking the skeleton is discussed in Part III (p. 170). If your understanding of sentence structure is still insecure, consult a good grammar book (see "Selected References").

English verbs and verb "look-alikes"

Verb forms (verbals) that look like verbs confuse many ESL writers. They are a particular problem for writers whose first language is Arabic. Sentences often contain one or more of these look-alikes—participles (verbs used as adjectives), gerunds (verbs used as nouns), or infinitives (which are used in many ways).

When a sentence includes one of these verbals, ESL writers are often satisfied that they have written a complete sentence. But if there is no real verb, the string of words is only a fragment of an idea. It does not make sense. It fails to make a statement. Certain types of sentence fragments are occasionally used by skillful writers for a special effect, but unintentional or accidental fragments create nonsense—garbled writing that cannot be understood.

To determine whether confusion about verbs and verbals is your problem, try Test 5 (Part III, p. 177). After checking your answers, you may be able to draw one of three conclusions:

(a) You are not yet ready to attempt long, complex sentences. Keep your sentences short. Keep them as simple as possible. Don't try to put too many ideas in one sentence. As mentioned earlier, you need not feel apologetic for brevity and simplicity. These are marks of readable writing—even for writers whose native language is English.

(b) You should continue formal study of English.

(c) You should read, read, read in English, preferably well-written material that interests you (e.g., *National Geographic* magazine. Also try *The Careful Writer* by Theodore M. Bernstein[9]; besides being interesting and instructive, it has a lot of sly wit.)

As you read, take time to study the way accomplished writers put their words together. Observe how they use modifiers, articles, and phrases. Keep a dictionary handy and use it every time you run into a new word; start enriching your vocabulary. Finally, practice picking skeletons.

Those troublesome verb tenses

Tense is defined simply as that verb characteristic which expresses time; that is, it tells **when** an action takes place, or **when** a state of being exists.

The trouble with these expressions is that they are so unstable. Their forms change not only with time, which is an easy concept to grasp; they also change with person, number, voice, and mode. To further complicate matters, we have special ways of using verbs (in all tenses) to express obligation or necessity, capability or capacity, ability or power, and so on. And to cap it all, we have irregular verbs whose forms make sense only to those able to trace the development of words through history to Modern English.

It is not surprising, then, that tenses often give many ESL writers headaches. Perhaps the greatest pain is suffered by people (Indonesians, for instance) whose own language is not burdened with verb tenses. At the request of many of these writers, we have devised a table to show how the tenses are formed (Table 6.)

The table is based on the assumption that the user is familiar with the *principal parts* of verbs, including irregular verbs:

Present: the infinitive without the *to*. For example: work,
draw, plan
Past Indicative: worked, drew, planned
Past Participle: worked, drawn, planned

If you have not mastered the principal parts of verbs, your English-language books have lists you can refer to. These forms must be memorized.

To review tenses briefly, we can represent the six tenses (three simple and three perfect) with a special time line that begins with the present and extends into either the past or the future:

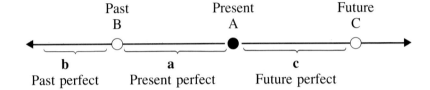

Past	Present	Future
B	A	C

b	**a**	**c**
Past perfect	Present perfect	Future perfect

The Simple Tenses

Past (B):   Expresses an action (or state of being) at a time before the present time.

Present (A): Expresses an action (or state of being) at the present time (now).

Future (C): Expresses an action (or state of being) at a time after the present time.

The Perfect Tenses

Past perfect (b):   Expresses an action (or state of being) that took place (or existed) prior to a time in the past.

Present perfect (a): Expresses an action (or state of being) that took place (or existed) prior to and up to the moment of speaking or writing. This is the tense we use to summarize status or progress: *We have completed five tests; five tests have been completed; the work has continued.*

Future perfect (c): Expresses an action that will take place (or a state of being that will exist) in the future, but prior to some future time.

Limitations of Table 6

We should point out that to keep this table a reasonable size, we did not include the progressive forms (the *-ing* forms) of verbs:

work: we are working; we had been working; we will be working; etc.

draw: They are drawing their own conclusions; they have been drawing their own conclusions; conclusions were being drawn.

Furthermore, Table 6 does not recognize idiomatic uses of tense, which must be learned through observation and experience. For instance, *I am leaving* is in the present progressive tense, but we use *am (are) leaving* idiomatically to express future time: *I am leaving tomorrow.*

Many ESL writers quite logically choose future tense auxiliaries (*will be*) to express future time in sentences beginning with *when* or *if: When* (or *if*) *this work will be done.* That is not correct English. To indicate future time in clauses beginning with *when* or *if,* use the auxiliary verb *is* (*am, are*) plus the past participle: *when* (or *if*) *this work is done.*

Nor does Table 6 show the auxiliary verbs that are used for special purposes. The chief ones can be summarized as follows:

(1) To express strong obligation or necessity, use *must, shall, have to, has to, had to*:
   We must (or have to) return.
   The rules shall be strictly adhered to.

(2) To express weak obligation or need, use *should* or *ought to*:
   We should rewrite the report.
   We ought to have written it long ago.

(3) To express ability or power, use *can:*
   He can use either method.
   They can be correlated.

(4) To express capability or capacity, use the present tense of the verb—no auxiliary is needed:
   That tank holds 150 liters.
   This equipment detects toxic gases.

TABLE 6A
How to Form the Simple Tenses

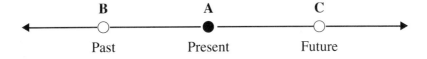

	B	A	C
	Past	Present	Future

With Active Verbs	*With Passive Verbs*

The Simple Present Tense (A)

Use the simple infinitive (the verb without the *to*).

I <u>use</u> this equipment.
We <u>draw</u> several conclusions.
This report <u>describes</u> a new technique.
This <u>is</u> a new theory.

Use *is* or *are* + past participle.

This equipment <u>is used</u>.
Several conclusions <u>are drawn</u>.
A new technique <u>is described</u> in this report.
(<u>Is</u> is an intransitive verb and cannot be made passive.)

The Simple Past Tense (B)

Use the past indicative:

I <u>used</u> this equipment.
We <u>drew</u> several conclusions.
That report <u>described</u> the new technique.
That <u>was</u> a new theory.

Use <u>was</u> or <u>were</u> + past participle:

This equipment <u>was used</u>.
Several conclusions <u>were drawn</u>.
The new technique <u>was described</u> in that report.
(<u>Was</u> is an intransitive verb and cannot be made passive.)

The Simple Future Tense (C)

Use <u>will</u> (or <u>shall</u>) + simple infinitive.

I <u>will</u> (or <u>shall</u>) use this equipment.
We <u>will</u> (or <u>shall</u>) draw several conclusions.
This report <u>will describe</u> the new technique.
This <u>will be</u> a new theory.

You <u>shall follow</u> the rules.

Use <u>will be</u> (or <u>shall be</u>) + past participle.

This equipment <u>will</u> (or shall) <u>be used</u>.
Several conclusions <u>will be drawn</u>.
The new technique <u>will be described</u> in this report.
(<u>Be</u> is intransitive and cannot be made passive.)

The rules <u>shall be followed</u>.

TABLE 6B
How to Form the Perfect Tenses

Past   *Present*   *Future*

b	**a**	**c**
Past perfect	Present perfect	Future perfect

With Active Verbs	*With Passive Verbs*

The Present Perfect Tense (a)
 Use <u>has</u> or <u>have</u> + past participle:
   I <u>have</u> <u>used</u> this equipment.
 	
   We <u>have</u> <u>drawn</u> several conclusions.
   He <u>has</u> <u>been</u> in Rome two weeks.
	Use <u>have</u> <u>been</u> or <u>has</u> <u>been</u> + past participle:
	   This equipment <u>has</u> <u>been</u> <u>used</u>.
	   Several conclusions <u>have</u> <u>been</u> <u>drawn</u>.
	   (<u>Been</u> is intransitive and cannot be made passive.)

The Past Perfect Tense (b)
 Use <u>had</u> + past participle.
 	
   I <u>had</u> <u>used</u> this equipment (before some time in the past).
   We <u>had</u> <u>drawn</u> several conclusions (before the project was finished).
   He <u>had</u> <u>been</u> in Rome two weeks (when he **decided** [past tense] to leave).
	Use <u>had</u> <u>been</u> + past participle.
	   This equipment <u>had</u> <u>been</u> <u>used</u>.
	
	   Several conclusions <u>had</u> <u>been</u> <u>drawn</u>.
	
	   (<u>Been</u> cannot be made passive.)

The Future Perfect Tense (c)
 Use <u>will</u> <u>have</u> + past participle:
   I <u>will</u> <u>have</u> <u>used</u> this equipment 10 years (by the time I discard it, in the future).
   When we complete this project we <u>will</u> <u>have</u> <u>drawn</u> some conclusions.
   He <u>will</u> <u>have</u> <u>been</u> in Rome two weeks next Saturday.
	Use <u>will</u> <u>have</u> <u>been</u> + past participle:
	   This equipment <u>will</u> <u>have</u> <u>been</u> <u>used</u> 10 years.
	
	   When this project is completed, some conclusions <u>will</u> <u>have</u> <u>been</u> <u>drawn</u>.
	   (<u>Been</u> cannot be made passive.)

(5) To express a general truth or principle, use the present tense without auxiliaries:

The moon <u>affects</u> the tides.

Honesty <u>pays</u>.

(6) To express a possibility or permission, use *may* or *might*:

It <u>might</u> (or <u>may</u>) rain tomorrow.

<u>Might</u> (or <u>May</u>) I be excused?

Yes, you <u>may</u> be excused.

(7) To express strong belief or strong probability, use *must be*:

The wiring <u>must be</u> faulty—this thing keeps shorting out.

(8) To emphasize, use *do* or *does*:

We <u>do</u> want our visitors to feel welcome.

It <u>does</u> appear that we are welcome.

(9) To express conditions that do not exist, use the subjunctive form:

If this well <u>were</u> a gusher (the plural <u>were</u> with the singular <u>well</u> says the well is not a gusher), I <u>would be</u> happy (<u>would be</u> says I am not happy).

If it <u>were</u> possible, she would help you.

You have probably noted that many usages can be learned from books, while others must be learned by close attention to usage by reputable writers and fluent speakers. It is often a matter of observing, listening, questioning, and gradually assimilating.

Use of *a*, *an*, and *the*

Like most European languages, English employs special adjectives called **articles.** The definite article is *the* and the indefinite article is *a*. Since *an* is only a form of *a* used for euphony, the choice between *a* and *an* is not a problem (see below). The problem is when to use *a*, when to use *the*, and when to use no article at all. This is really troublesome for people whose first language has no articles (e.g., Chinese), because the rules are hard to explain; native English speakers learn from infancy, by listening and reading, without the use of rules.

Use *a* before **consonant sounds**, regardless of spelling: *a home*, *a unit*, *a yolk*. Use *an* before **vowel sounds**, regardless of spelling: *an honest mistake*, *an X-ray*, *an yttrium compound*.

Here are some general guidelines on the use of *the*:

(1) *The* specifies a particular person, place or thing:
   The writer who prepared the memorandum
   Among the attractions in the state

(2) *The* identifies a person, place, or thing introduced previously with *a* or *an*:
   The yttrium compound . . . (referred to above)
   The coating of plastic . . . (mentioned above)

(3) *The* converts a common noun or adjective into a generic term (a class) or a typical specimen of a class:
   The pen is mightier than the sword.
   The rich and the poor support this candidate.
   The cheetah is faster than the greyhound.

(4) *The* precedes a title (denoting rank or office) of a person who is not called by name:
   The governor notified the senator.

(5) *The* distinguishes one of a class from others in the same class:
   These pies are all good, but I prefer the chocolate.
   Americans speak English, but the southerner does not sound like the northerner.

(6) *The* is dropped in front of some nouns when function is referred to:
   They went to school. (A general statement meaning they attended school. ''They went to the school'' would mean a particular school building—possibly for a purpose other than education.)

(7) *The* is usually not used before nouns that are definite without it:
   London (no the) is the largest city in England (no the).
   (But ''The London I'm talking about is in Ontario.'')

(8) *The* is not used before classes of things that cannot be divided into individuals:
   Cool water (no the) refreshes.

(9) When two words refer to the same person, place or thing, *the* is not repeated:
   The president and (no the) chief executive officer is Sandra Perkins.

Studying daily what you read in English will help you master *a* and *the*. Use a newspaper, magazine, or journal that you enjoy.

For starters, do this only for a couple of pages per day. Stop to examine how *the, a,* and *an* are used. This practice will give you something that rules cannot: a feeling for uses that will guide you unconsciously as you write. Then the foregoing guidelines will be useful in editing.

Sequence of adjectives in a series

When you use more than one adjective in front of a noun, the order in which they are placed usually matters, although the sequence is admittedly unsupported by any obvious logic. Deviations from the general sequence can sound very strange or amusing to English-speaking readers. The general sequence of adjectives is as follows:

(1) Begin with articles (*a, the*) or adjectives such as *our, their, my, your, his, her; this, that, these, those; which, what, whose; no, none, some, several, many; such, enough; either, neither.*

(2) Next, the cardinal numbers: *one, 2,000, dozen.*

    (1) (2)
The five conclusions drawn from this work . . .

The ordinal numerals (first, sixth) may come either before it or after other adjectives, depending on the meaning:

    (1) (2)
The first model . . .

       (1)   (2)
But: His inefficient first model is not at all

       (1)   (2)
like his efficient second model.

(3) Judgmental terms (based on opinion or perception): *beautiful, expensive, efficient, ancient, cold.*

    (1) (2)   (3)
A second , more efficient model . . .

(4) Size terms: *large, big, 5-ton, 50-mm, 20-acre.*

    (1) (3) (4)
A beautiful big apple

    (1) (3)   (4)
An efficient 5-ton air conditioner

(5) Shape terms: *square, spherical, L-shaped, flat.*

    (1)   (4)   (5)

    <u>A</u> <u>large</u>, <u>L-shaped</u> building . . .

(6) Color terms: *red, greenish, rosy-hued, multicolored.*

    (1)   (3)   (4)   (5)   (6)

    <u>A</u> <u>beautiful</u> <u>big</u>, <u>round</u>, <u>red</u> apple . . .

(7) Other terms and sequences: Sometimes the order depends on what "sounds good"—what is most euphonious. In the following examples the author might think one of the sequences "sounds better" than the other.

    Our big old rickety, ramshackle house.
    Our big, rickety, ramshackle old house

A series of adjectives is often interrupted by adverbs such as <u>very</u>, <u>more</u>, and <u>most</u>, as in:

    (1)   (3)

    "<u>A</u> **very** <u>important</u> concept."

Such an adverb must always be placed before the adjective it modifies.

The punctuation of adjectives in a series is also a matter of judgment (or idiom) in many cases. Here is a rather complex, generalized rule that may be helpful:

If the first adjective seems to modify all the rest of the phrase (the other adjectives plus the noun), a comma may be omitted. For instance, in this expression

A big old rickety, ramshackle house

the terms *big* and *old* may be felt to modify "rickety, ramshackle house." Thus the writer might choose not to use commas after *big* and *old*. But in a sentence like this

This is an efficient, inexpensive, updated model.

the author sees each adjective as modifying *model* only. Therefore commas would be used after all adjectives except the last one.

In summary, rules are not infallible guides. To write confidently, placing adjectives in idiomatic sequence, requires a feel for English that you can best obtain by observant reading.

Placement of adjectives and participles

You have probably noticed that while adjectives and participles (verbs used as adjectives) commonly precede the noun, they may also follow it:

The *large,* evenly *distributed* pores are . . . or
The pores, *large* and evenly *distributed,* are . . .

In the first phrase, *large* and *distributed* are essential to the description of pores.

In the second phase, *large* and *distributed* are shown as non-essential—just additional information. Commas are always used to set off nonessential modifiers. The phrase is actually part of a clause in which the words "which are" or "that are" are understood:

The pores, [which are] large and evenly distributed, are . . .

Here are further examples:

(1) Essential adjectives or participles (no commas):
Challenges (that are) big enough to interest these investigators are hard to find.
A scientist (who is) experienced in grantsmanship can write good proposals.
Technicians (who are) working in this lab must observe safety precautions.

(2) Nonessential adjectives or participles (commas used):
These challenges, (which are) big enough to interest most investigators, arise frequently.
Scientists, (who are) often experienced in grantsmanship, find it easier to write good proposals.

Placement of adverbs

In simple sentences, adverbs ending in *-ly* can be moved around quite freely, depending on the emphasis desired:

They walked home slowly.
They walked slowly home.
They slowly walked home.
Slowly they walked home.

But adverbs are not usually placed between a verb and its object:

This: They formed new ideas quickly.
Or this: They quickly formed new ideas.
Or this: Quickly they formed new ideas.
But not this: They formed quickly new ideas.

The trouble with learning a language is that all rules have exceptions. For example, when some infinitives are the object of a verb, we usually place the adverb between verb and object:

She wanted very much to visit India.
He decided right then to go.

The second sentence clarifies the reason for this exception. If the adverbs *right then* followed the object (the infinitive *to go*), the meaning would be different: He would not be "deciding right then" about a future trip but would be deciding "to go right then" (immediately).

Double negatives

English follows the logic of mathematics in this respect: two negatives make a positive and cancel each other out. In some languages two negative words can be used in the same sentence for emphasis, but in English this is a vulgarism associated with the poorly educated.

In watching out for this problem, remember that there are many negatives besides *no, not,* and *nothing.* You must also look out for *never, neither, none, nobody, hardly,* and *scarcely.*

Here are some bad examples to avoid:

We do not need no more spare parts.

The chance of fraud was scarcely considered neither.

We couldn't get results with none of the equipment.

They checked the instruments but did not find nothing.

The worst cases are probably those in which the two or more negatives are directly adjacent:

There was not hardly no improvement in cash flow.

<u>Nobody</u> <u>never</u> told me this was poor English before.

As usual, there are exceptions. Since two negatives make a positive in English, they are acceptable in a few constructions designed to create a very weak affirmative statement:

Idiomatic English appears difficult, but <u>not</u> <u>im</u>possible.

I am <u>not</u> entirely <u>un</u>grateful for this help with idioms.

Misused infinitives

Following the verbs *recommend, suggest, insist, request,* and *require,* idiom calls for a noun, gerund, or subjunctive construction—not the infinitive that would be used in many other languages. In English, the infinitive sounds awkward, strange, even incorrect:

Not This	**But This**	**Or This**
We **recommend** <u>to</u> <u>hold</u> the conference in Paris.	We **recommend** holding the conference in Paris.	We **recommend** **that** the conference <u>be</u> <u>held</u> in Paris.
She **requests** <u>to</u> <u>ap</u>prove an expenditure of $1MM.	She **requests** <u>ap</u>proval of an expenditure of $1MM.	She **requests** **that** an expenditure of $1MM <u>be approved</u>.
They **suggest** to <u>study</u> this plan further.	They **suggest** further <u>study</u> of this plan.	They **suggest** **that** this plan <u>be studied</u> further.
He **insists** <u>to</u> <u>have</u> an independent audit.	He **insists** on <u>having</u> an independent audit.	He **insists** **that** an independent audit <u>be carried out</u>.
They **require** <u>to</u> <u>have</u> all applicants take a medical examination.	They **require** a medical <u>examination</u> for every applicant.	They **require** **that** all applicants <u>take</u> a medical exam.

False friends

When you are building your vocabulary in English, be careful of "false friends." A false friend is a word in English that **looks** like a word in your own language but which has, or can have, a different (sometimes very different) meaning. This is a problem mainly for speakers of Germanic and Romance languages, who will find many false friends in English.

Here are two examples. In Spanish and French, the primary meaning of *actual (actuel)* is "at the present time." In English, the primary meaning is "real; existing; factual." In German, *also* means "thus; accordingly"; in English it means "besides; in addition."

Useful References

For serious students of the English language, we particularly recommend the following books (see "Selected References," p. 255):

The Careful Writer[9] by Theodore Bernstein. This book is especially good for ESLs, with help on which prepositions follow certain nouns, verbs, and adjectives.

A Dictionary of Contemporary American Usage[5] by Bergen and Cornelia Evans

A Dictionary of Modern English Usage[6] by H.W. Fowler.

Descriptive English Grammar[3] by Homer House and Susan Harman.

Part V

Writing in an Era of Rapid Change

"There is nothing permanent except change."
— *Heraclitus (513 B.C.)*

Part V
Writing in an Era of Rapid Change

> Words long faded may again survive;
> And words may fade now blooming and alive
>
> —Horace, *Ars Poetica*

The social, technological, and economic changes introduced during the twentieth century have changed the way we work and thus the way we think, speak, and write on the job.

In the last few decades particularly, writers have been forced to deal with new language ideas. Computer science, in introducing a new tool, has introduced a whole new idiom. Changing social customs and moral standards, as well as an invigorated concern for minority groups, have brought about other new vocabularies that are controversial and that will continue to be so until standards have been stabilized. Words may indeed "fade now blooming and alive."

This section is concerned with the effects of those changes on the way we compose and produce technical, scientific, and business documents.

WRITING IN THE AGE OF THE COMPUTER

In writing about computer applications, one of the few generalizations that will hold up is this: by next year (if not next month or even next week) the situation will be different. Statements and opinions about electronic data processing tend to have a very short shelf life—typical of the environment we refer to as "an Era of Rapid Change," of which the computer is both an example and an agent.

Text management—called "word processing" (WP) to distinguish it from earlier kinds of electronic data processing by computer (EDP)—has long been available for big (mainframe) computers and "dedicated" word processing equipment. Word processing is, for the moment, the single most popular use for microcomputers. It has not, however, ushered in a millennium of reduced paper work. Computers have indeed made it easier to file and retrieve and to handle data and documents, but so far they have not reduced the blizzard of paper that most of us have to contend with. If anything, WP with high-speed printers has, like

237

the ubiquitous photocopier, made it all too easy to churn out even more "hard copy."

The other factor that has made WP a mixed blessing for more than a few people has been the software industry's spotty record in developing programs and instruction manuals that are truly "user friendly." This topic is the subject of a subsequent section.

Word Processing as an Aid to Writers

We mentioned earlier the convenience of "talking" into the keyboard of a word processor so that words, sentences, paragraphs and pages can be manipulated electronically in just about every way they have been treated manually in the past. Text management via computer or dedicated word processor may never totally replace the ordinary typewriter, any more than the typewriter has totally replaced pens and pencils. But the increasing power and flexibility of computers—the growing number of ways in which words can be not only handled in the computer but also entered into it and sent out from it—ensures that WP will effectively replace the typewriter in most businesses and in many other situations as well.

For readers who may not be familiar with WP capabilities, here are eight examples:

(1) The basic, and best-known, capabilities in word processing involve the speed and ease of editing and formatting. Text can be added, deleted, transposed, selectively replaced, retrieved from an electronic "file," shifted around in chunks large or small, and then stored. It can be "formatted" (a new verb, scarcely needed until WP came along) by instructing the computer as to pagination, margins, line spacing, "pitch," centering, and justification.

(2) Access to a "database" file of names and addresses, in conjunction with the editing capabilities just mentioned, permits the WP user to create customized or personalized form letters for mass mailings—including the mailing labels.

(3) As the price of computer memory and processing speed continues to drop, other editing functions become available or more powerful. The simplest of these is probably the automated spelling checker and hyphenator, which requires a dictionary, or wordlist, of correct spellings and hyphena-

tions in the computer's memory. Such a program finds (and flags for the user) every word that is not on its list. The user may then correct a misspelling or, if the word was correctly spelled but not on the list, add the new word to the list.

(4) Similarly, there are programs for checking a text against a stored list of clichés, commonly misused words and phrases, passive verbs, vague terms, trite or awkward phrases, and "sexist" terms in order to flag any such occurrences. Some programs will also suggest substitutes (a thesaurus function). A full-size thesaurus is also available for direct access during text processing.

(5) The power of WP programs to detect grammatical errors or nonstandard usage is limited only by computer memory size and the skill of the programmer. We can anticipate major advances in this category; they may even help arrest the decline of standard English.

(6) A final example of editorial help from the computer is the automatic calculation of readability indices mentioned on page 131. Again, the precision of such calculations depends on available memory size, which determines to what degree the program can recognize and count (*a*) syllables, (*b*) phrases to be treated as a single word, and (*c*) ends of sentences, as distinct from periods used within a sentence.

(7) Text, tables, and graphics, including an endless array of type faces, can be created together, arranged on a single page, and printed as a unit by the author at the keyboard— a spectacular advance that we will soon take for granted.

(8) Automated translation from one language to another is yet another area where capabilities are limited only by computer memory size and the patience and ingenuity of the programmer. Rote word-for-word substitution is of course child's play for a computer, but idiomatic translation requires that entire phrases be recognized and handled properly.

What Next for Computerized Writing?

In the short term we can of course anticipate greater power and flexibility in all of the eight operations described above. As computer memories get larger and processing times get shorter, software writers will stretch their inventiveness to make use of these advances.

Voice recognition, or speech recognition, mentioned in Step 7 (Dictate) may well be the next major advance in word processing. Programs and hardware that enable computers to respond to certain spoken commands already exist; but as this book is being written, we have not yet reached the stage where anyone can routinely enter text into a computer by talking to it as one would to a stenographer, thus bypassing the keyboard entirely.

When this capability arrives, it will start another revolution in business communication: dictation to a computer will become the standard way to create first drafts (although some people might still end up using the keyboard to correct those drafts). Dictation will then be a required skill in cost-conscious organizations.

Practical applications of "knowledge processing," described as the next logical step beyond data processing, may be slow in reaching the marketplace for a while. But when such applications become common, we can be certain that some will have profound effects on the way we take care of our scientific, technical, and business communication by "hard copy." At the present time, it would be idle to speculate about what such effects might be.

MAKING MANUALS
"USER FRIENDLY"

The News

Use the 25 suggestions in this section to apply the Murray System to the preparation of instruction manuals—efficient help in easy-to-use form.

Long before the buzzword "user friendly" arrived on the scene, people were throwing instruction manuals across the room with howls of frustration. The old adage "When all else fails, read the instructions" is not much help if the instructions seem written in pidgin Greek. But the critical importance of easy-to-follow procedures has been spotlighted lately by intense competition among the writers of computer software.

These programmers realized they could sell software as a way of making hardware user friendly. But they were slower to realize that good "documentation" (computerese for "instruction manual") is equally vital to making the **software itself** more friendly. Some of the most impenetrable prose written in English today is churned out by computer specialists who know their subject, but who haven't learned how to communicate that knowledge.

While computer documentation provides a glaring example of the user's plight, all kinds of written instructions produce frustration in readers. This is true of the quaint quasi-English that accompanies some products made in the Far East. It is also true of the dense, cryptic technologese written by native English speakers who seem to be trying to dazzle readers with obscure vocabularies. Such incomprehensible instructions can be dangerous. Consider the following from an industrial safety manual:

> In order to help safeguard individuals, "Threshhold Limit Values" have been established by the American Conference of Government Industrial Hygienists which sets forth the maximum permissible atmospheric concentrations of contaminants, which may not be exceeded if a safe working environment is to be guaranteed.

In the event of a toxic gas leak, how many lives would this "warning" save? It doesn't even say what the safe limits for specific contaminants are. How much more effective to instruct:

IF ATMOSPHERIC CONCENTRATION OF X GAS REACHES _____, EVACUATE THIS WORK AREA IMMEDIATELY.

Nobody claimed that writing a good set of instructions was easy. One reason manuals are notorious for their poor writing is that they are among the most difficult writing jobs that technical and business people face.

Although authors of manuals may not be trained as writers, they have to describe long, complex procedures involving highly sophisticated technology—in a way that will be clear to all readers, including those who are totally unfamiliar with the technology. Manual writers may have to meet stringent safety requirements of their institution. They may have to satisfy both the readers who want the detailed "how-to" and the bosses who criticize them for getting bogged down in detail. And often they are not given enough time for the writting job.

To make matters worse, some technical hot-shots and scientists regard ordinary people with unspoken contempt (see "Real Programmers"). Their attitude is "if they can't understand it, they shouldn't be reading it." Manual writers who feel that way simply cannot be bothered to consider the reader.

"Real" Programmers

Real programmers don't write specs—users should consider themselves lucky to get any programs at all, and to take what they get.

Real programmers don't comment their code. If it was hard to write, it should be hard to understand.

Real programmers don't write applications programs; they program right down on the bare metal. Applications programming is for feebs who can't do system programming.

Real programmers don't document. Documentation is for simps who can't read the listings or the object deck.

—Excerpted from "Real Programmers"
in the Newsletter of the
Computer Oriented Geological Society,
December 1984

Nevertheless, the writing of a user-friendly manual must start with consideration of the ultimate user—not just any user, but the least experienced one. If the novice can understand the instructions, other users should have no problems.

Our 25 suggestions for applying Engineered Writing techniques to the preparation of a manual start, of course, with the user.

Begin with the User

(1) Analyze your people. Who are the users? What do they need to know? How they will use the equipment or procedure about which you are instructing them? How much or how little do they know about it already?

(2) If you do not know your audience, find out. Ask questions.

It may cost you time and trouble, but you neglect this preparation at your own risk.

(3) As you write, put yourself in the reader's position, following your own instructions. Anticipate problems (remember Murphy's Law). And anticipate the reader's "what if . . . ?" questions.

Organize Your Message

(4) This is Step Two: Develop Your Essential Message, beginning with your overall NEWS. That will usually be an expansion of the manual's title: "Here is how you do such-and-such."

Then answer any WHY questions. (Sometimes a supervisor wants to include here a lot of background or historical information. Try putting it in an appendix.) The opening summary should be reserved for a broad overview of the manual, and the introduction should be reserved for information of real use to the manual user. (Of course, if history or background would help readers get the most out of the manual, the introduction would be a good place to include it.)

Now summarize HOW. For manuals, HOW can usually be summarized only in broad general terms (e.g., the sequence of major steps in a procedure, without detail; the objectives of those major steps; or significant changes from previous procedures).

(5) When you have completed a topic outline showing all major operations or steps, you can anticipate what your readers' questions will be at each step: WHAT do I do? Perhaps WHY do I do that? Certainly HOW (step-by-step) do I do it? And maybe NOW WHAT? (What will be the result? What options, if any, do I have as the next step?)

(6) At the beginning of each major unit within your procedures, tell the readers where they are headed; give them a brief orientation. If appropriate, include a flow diagram like that in Fig. 2 (p. 20) to show the overall steps. This is especially helpful if you have any branching in your procedure, with a separate subsequence of instructions to be followed for each different situation that can be encountered.

How to Give Instructions

(7) Use the imperative (command) form of the verb for instructions, as we have done here.

(8) Put your instructions in the plainest, simplest words you can. Avoid high-tech jargon as much as possible. Every technical term must be clear to your readers. Where technical jargon must be used, include at the beginning of the text unit a list of all the terms in that unit, with the plainest definitions possible, even with small pictures or drawings alongside the definitions where appropriate. If there aren't too many definitions, try putting them in a box, or highlight them in color. In any case, provide a complete glossary of technical terms at the end of the manual.

(9) Make sure your instructions are given in the sequence in which they have to be carried out (where it doesn't matter, say so—unless that is obvious).

(10) Break instructions down into small, logical steps.

(11) Number the steps to emphasize sequence, to make it easier for readers to find their way around, and to allow you to refer conveniently to different steps.

(12) Consider your readers' technical background and needs before deciding the most effective way to describe complicated equipment. Readers may best be served by—
   ■ Proceeding from the simple to the complex:
      The entire apparatus, which is complicated, can best be described by beginning with its most basic element, the heater (see diagram A).
   ■ Describing by function:
      An electric heater (A), which maintains the temperature in cell (B), is controlled by device (C).
   ■ Describing from the inside out:
      The core of the model is the heater, which is placed in an oil-filled cell. All this is enclosed in an 8-inch-diameter housing.
   ■ Describing from the outside in:
      The system has an 8-inch-diameter housing in which is centered an oil-filled cell containing the heater core.

(13) Do not write merely to be understood; write so that you can**not** be **mis**understood.

(14) Never try to impress readers with your technical know-how; it is much better to impress them with your ability to prepare a user-friendly manual.

How to Help Users Find Their Way Around

(15) Include a complete table of contents and complete lists of tables and figures, backed (if the manual is a long one) by a comprehensive subject index.

(16) Use plenty of clear, bold headings and subheadings throughout. They are a great time saver for readers who are already somewhat familiar with the procedures and thus don't need to start at the beginning and read every word. Consider using informative headings (those containing a verb) where appropriate.

(17) If your manual is a substantial one, consider using index tabs to mark the major subdivisions.

Other Aids to User-Friendliness

(18) Use pictures freely. Clear, accurate, and carefully labelled line drawings are often preferable to photographs, in which the reader may or may not be able to see what is obvious to you, the expert.

---

(19) Put safety information (and other important instructions) in boxes like this one, but don't overdo this sort of highlighting.

---

(20) Use white space (and color, if it's affordable) to improve readability.

(21) If you think readers would ask WHY a particular instruction must be followed, don't hesitate to explain—especially when there's a safety reason for carrying out some apparently arbitrary step.

(22) If you must include a lot of highly specialized information that is not part of the main sequence of instructions, con-

sider putting it in an appendix. (But be sure to refer to that appendix where appropriate in the instructions.)

(23) If your manual will be revised periodically, include a request for feedback from the users.

Now What?

(24) Edit rigorously. Check your verbs. Active, command-form verbs will make instructions clear and easy to follow correctly. And they will reduce length by as much as one-third.

(25) For a final check, give a draft copy to a couple of nonexperts and have them try to follow your instructions. Such a "dry run" can prove surprisingly helpful in finding items that need clarifying.

SEXISM IN ON-THE-JOB WRITING

We need to distinguish three kinds of "sexist" terms in English:

(1) Deliberately offensive usage, which by definition is avoidable;

(2) Usage that is built into the language but that can be conveniently avoided through the use of other words;

(3) Usage that is built into the language and *cannot* be conveniently avoided.

For the first two kinds of usage, the sensible course of action is obvious: avoid them. To do so is an act of courtesy and thoughtfulness. If you want the goodwill and cooperation of your readers, why antagonize some of them needlessly?

For the remainder of this section, we can focus on the third kind of language—sexist terms that are ingrained in English and cannot be avoided without clumsy or ungrammatical circumlocutions. It would be nice if writers could find a convenient route through this new minefield without offending anybody. In some situations, we can; in others we cannot find a totally inoffensive route that avoids doing violence to the language.

He/She/They; Him/Her/Them; His/Her/Their

Like many other languages, English has no genderless, third-person pronoun in the singular (apart from *it*). With genderless singular antecedents like *everyone, somebody,* and *each,* English solves the problem by using forms of the personal pronoun *he,* as in "Nobody is going to get his way all the time."

From the standpoint of linguistic history, *he* does have some claim to be the compromise pronoun of choice when either or both sexes are referred to. In Old English the singular pronoun for the female was *heo* (for the male, *he*). *Heo* subsequently dropped out of use, to be replaced by *seo,* which later gave way to *she.*

In the last decade or so, efforts to avoid *he, him,* and *his* have centered on three replacements:

(1) *he or she; him or her; he/she; his/her;*

(2) *s/he* **(no equivalent seems to have been dreamed up for** *him/her* **or** *his/her***);**

(3) *they* **(along with** *them* **and** *their***).**

The first replacement (*he* or *she*) is grammatically sound, and there are times when it is essential for clear meaning. But it is often a cumbersome circumlocution.

> Any writer who forces himself or herself [or "herself or himself"] to change her or his writing in order to prove that his or her attitudes are nonsexist will find herself or himself in trouble when he or she discusses people.

The easy solution is to use the generic *he.* If this offends, you can sometimes pluralize: "Any writer who forces himself . . ." could be rewritten "Writers who force themselves . . . ."

The second replacement, *s/he,* is an unpronounceable gimmick that has no place in educated writing.

The third replacement (*they, them, their* with singular nouns and often with singular verbs) is an egregious grammatical blunder. There is no kinder way to describe it. There is no justification for its use.

Conclusion: If *it* and *its* are inappropriate, use *he, him,* or *his* as the genderless singular pronouns. Or find a way to recast the sentence, for example by pluralizing. **But never imagine that a plural pronoun can agree with a singular antecedent**.

It's Every Person for Theirself

However often errors like these may be committed in casual speech, there is no excuse for them in writing:

"In the long run, each participant gets what they deserve."
"Anyone interested in nominating themself or someone else . . ."
"A true artist has their own way of doing it."
"The patient pictures themselves in a relaxing situation . . . ."
"Everyone has it in them to succeed."
"A person goes up when their name is called and chooses theirs . . ."

These examples of efforts to avoid sexist usage were culled from current sources (omitted to protect the guilty). How would you recast these sentences?

Ms.—A Useful Invention

Ms. is a very handy invention that will undoubtedly gain increasing currency in English, because it fills a need. We don't know or care about the marital status of the women we write to on the job.

On the other hand, if a woman signs herself as *Mrs.* or *Miss,* she evidently prefers the term to *Ms.* Courtesy in addressing her calls for honoring that preference.

It is interesting to note that most European languages have exact equivalents of *Mr., Mrs.,* and *Miss.* (Like English, they lack a separate term to show that a male is unmarried.) But they have yet to develop terms analogous to *Ms.*

-man, -person

The common suffix *-man* has come under the same sort of attack as *he/him/his.* The replacement that has been suggested is *-person.* Here are two relevant observations:

In old English, the word *man* did not necessarily mean the male of the species, but a person or human being. [It still

De-Manning The Language

What are your options if you consider the terms on the left "sexist"?

<u>   Some Suggestions   </u>

middleman	intermediary, distributor, wholesaler (not *middle-person*
postman, mailman	letter carrier, mail carrier (not *postperson* or *mail person*)
manpower	human resources, personnel
policeman	police officer, constable
cattleman	rancher
mankind	human race, humanity
housewife	homemaker
foreman	supervisor, crew chief
man-made	hand-made, machine-made, artificial, synthetic

retains this meaning in modern German—Ed.] So using the title *chairman* is perfectly acceptable for everyone.*

Now Therefore, be it Resolved, that organizations and parliamentarians of the National Association of Parliamentarians must use the term "Chairman" instead of "Chairperson" . . . .†

It may be noted in passing that the National Association of Parliamentarians has prominent women members.

Some people, perhaps thinking that *chairperson* has an unfortunate parallelism with *frogperson* (for scuba diver, or perhaps a being from another planet), have taken to calling themselves *chairs*. This solves the problem about as well as calling a frogman a *frog*.

---

*Della A. Whittaker, "Speak with Sense, Not Sexism," *The Toastmaster Magazine*, Feb. 1977, p. 12.
†National Association of Parliamentarians, Twentieth Annual Convention, San Francisco, 1980.

Certain other terms like *spokesperson,* although they may sound awkward, will undoubtedly gain currency as people become accustomed to them. And there are imaginative terms, like *fire-fighter* to replace *fireman,* which avoids risking the ridicule merited by *fireperson.* Other neologisms like *drafter* (to replace *draftsman*) and *waitperson* to replace *waiter* and *waitress* (seen on a menu in Santa Fe) seem less inspired and may not survive (we hope).

In summary, we recommend reasonable efforts by writers to avoid giving offense with terms that may be considered sexist. The operative word is "reasonable": there is no need to overhaul the mother tongue (not *parent tongue*) in a laudable but misguided attempt to remove imagined handicaps.

ETHNIC ENGLISH

The various types of ethnic English are neither appropriate nor effective for use in business, scientific, or technical speech or writing. Within an ethnic group, the style of social discourse is entirely the business of that group, but **the ability to shift from ethnic to fluent standard English as the occasion requires is essential.**

WHY? Because

(1) Standard English is the language used, understood, and accepted by a majority of educated readers, colleagues, and employers;

(2) Nonstandard language is as inappropriate in the board room or in a scientific report as jeans and T-shirt would be at a White House dinner; and

(3) Rightly or wrongly, fairly or unfairly, *non*standard English is generally viewed as *sub*standard English, its use classifying one in a way that will limit career advancement. As columnist William Raspberry pointed out in what he termed "one more piece of unsolicited counsel for minority students": "Proper use of the language is routinely accepted as a mark of intelligence, the first basis on which we are judged by those whose judgments matter."*

---

*William Raspberry, 1985, "Language Skills Can Open Doors," St. Louis *Post-Dispatch,* Feb. 12, 1985, p. 3B.

The different types of ethnic English include vivid idioms with inimitable ways of expressing ethnic humor. Modern popular music relies on them. But it does not follow that nonstandard English is equally suitable for scientific, technical, and business writing or speech.

Users of ethnic English, backed by some sociolinguistic experts, question whether one should abandon the right to his native way of speaking, in which he takes cultural pride. That question is not relevant to the matter at hand. An idiom's legitimacy and the rights of its users unfortunately have no bearing on its utility in putting across ideas. For our purposes, we can assume that ethnic English dialects are indeed legitimate dialects. Moreover, we must point out that everybody has the right to use any kind of language that pleases him, regardless of consequences. But the question we have to focus on is **what kind of language best meets the needs of both reader and writer, both speaker and listener.**

At the present time, there are still standards for English grammar and usage below which we descend at our professional peril. They set the level at which we speak and write so as to communicate without causing misunderstanding, distraction, or rejection. That kind of communication is our objective. Anybody who encourages us to consider any nonstandard English as acceptable in a business letter or scientific meeting displays arrogant disregard for our professional welfare.

It's ironic that the language "experts" who offer such encouragement do so in very clean and careful standard English. How could language experts do less? They are more aware than most that unfamiliar idioms call attention to words instead of sentences. Ungrammatical sentences call attention to the writer or speaker and his level of literacy instead of his ideas. Incorrect pronunciation and ethnic slang divert attention from the message to its author's background and education. All these false signals can bias perception of the writer's professional competence.

In short, the scientist, business person, or technical professional who uses an ethnic idiom because "I have a right to" is simply handicapping himself.

ORGANIZING A SPEECH

Speeches differ in their organization from written documents. A skillful lecturer, for instance, will use opening remarks—humorous, if he is good at jokes or story-telling—that give the audience time to settle down.

But the purpose of technical or business talks is not primarily to entertain, although anecdotes, analogies, and metaphors certainly lighten a subject. The purpose is to inform or educate. Thus it behooves us to get to the point after the opening remarks and then support that point. And at the end of the talk, since listeners usually do not have a copy of the speaker's text to refer to, the speech—unlike short written documents—should summarize the NEWS again.

Here is one big difference from a written presentation:

It is true that a speaker is free to impose the Suspense Format on his listeners, because they have no way of skipping ahead to the end of the message, looking for buried news. Indeed, the Suspense Format is essential in telling jokes, ghost stories, and anecdotes with a punchline. But it is a major mistake to hold back the news in an informational talk. If the listeners do not get bored and stop paying attention, they may at least get confused and lose the thread of what the speaker is saying. To make sure that a technical or business message gets through, then, base your speech outline on (1) the Analysis of People and (2) the guide questions of the Essential Message.

Cues for Nervous Speakers

If you are apprehensive about giving an oral presentation, the Expanded Outline (Step 5, p. 97) makes a fine security blanket to have at the lectern. Whether on full-size pages or a series of index cards, the outline is a good compromise between a few cryptic clues scribbled on a piece of paper and the complete text of a talk. (The complete text is a bad idea, anyway. You may lose your place if you don't read it verbatim; if you do, you lose eye contact, you forget gestures, and your voice is more likely to level off to a monotone. Reading is a time-tested way to put children to sleep, and it works with adult audiences, too.)

Make sure that your Expanded Outline has (1) headings underlined or highlighted in color and (2) marginal notations of slides or overhead transparencies where appropriate.

Visual Aids

We ordinarily read printed material at a distance of 12 to 14 inches (30 to 35 cm), but the back row of an audience may be 65 feet (20 meters) from a screen that is only 4 to 6 feet (120 to 180 cm) wide. Viewers have limited time to study a slide or overhead transparency; add poor projection equipment or inadequate darkening of the room, and we have a real problem of legibility.

The following techniques will help ensure readable and attractive slides and transparencies:

(1) Use bold lines, large print, and plenty of white space between lines of print.

Make illustrations simple, bold, and clear. Use heavy lines. Keep illustrations uncluttered. Here's a good test: If you can hold up a 35mm slide and read all the words on it with the unaided eye, it will be okay when projected.

(2) Limit each illustration to one theme.

Each slide, overhead transparency, or flip-chart sheet should introduce only one main idea. This will often mean that you must break up a complex illustration into several simpler ones. However, by starting with a basic framework (such as a base map or a very simple graph) and successively adding information in each illustration of a sequence of related illustrations on the same base, you can build up to a fairly complex picture without confusing your viewers.

(3) Take advantage of color.

Expensive for report illustrations, color is cheap for visual aids. Avoid overuse of black-on-white for projected illustrations; it's hard on the eyes. White on dark blue is eye-appealing and easy to read; so also is black on yellow or on other light colors. Red is good for accenting but not for letters or backgrounds. Black on dark blue is almost impossible to read.

(4) Use illustrations to cue your talk.

If you have enough illustrations, you can use them in place of cue cards (either type of cuing is preferable to putting your audience to sleep by reading to them). You can help your audience—and yourself—by including some "transition cues" on title slides (or overhead transparencies) that announce the name of the next topic to be covered.

GREATER INFORMALITY
IN WRITING AND SPEAKING

Ever since English reached a peak of formality and pomposity during the Victorian era, our language, both written and spoken, has tended to become increasingly informal. And in the past decade or so, this trend has clearly accelerated.

Thoughtful people wonder whether to accept the trend or resist it. Rather than go "whole hog" one way or the other, we think it is prudent to take a more selective attitude, at least in the areas of scientific, technical, and business writing:

- To the extent that it encourages simpler, more direct, and more readable writing, informality can be welcomed; but

- Where it leads to solecisms, barbarisms, and an impoverished vocabulary, the trend should be condemned for the "language-trashing" that it often becomes.

The English language is not only a central part of our inheritance as a people; it is also one of our most important tools in daily living. Properly used, it is a precision instrument. What would we think of a brain surgeon who operated with a can opener or rusty jackknife instead of a scalpel or laser beam? What would we say of a carpenter who used a screwdriver to drive nails, or a chisel to put in screws?

Yet this is the way some people treat our primary and most valuable instrument of communication. And since the language is the common property of all its users, those who misuse it damage it—indirectly—for all of us.

Selected References _____

BASIC REFERENCE WORKS

1. ***American Heritage Dictionary**   Edited by William Morris
   Houghton Mifflin Co., Boston   International Edition, 1979
   This dictionary is one of the few to include usage notes. Unlike Webster's
   third edition, it takes the responsibility for indicating preferred spellings,
   pronunciations, and senses of use.

2. ***Roget's International Thesaurus**   Revised by Robert L. Chapman
   Thomas Y. Crowell Co., New York   Fourth Edition, 1977
   The latest editions of this most useful of all word-finders are remarkably
   inclusive. Some writers prefer the dictionary format of this work—in essence, a dictionary of synonyms. Reduced paperback versions and diskette
   (computer software) versions are available.

3. **Descriptive English Grammar**   Homer C. House and Susan E. Harman
   Prentice-Hall Inc., Englewood Cliffs, NJ   Second Edition, 1950
   A thorough analysis of English grammar. The second half of the book uses
   diagramming to describe the structure of English sentences. Use this book
   when you want to get back to the basics.

4. **Harbrace College Handbook**   John C. Hodges and Mary E. Whitten
   Harcourt Brace Jovanovich, New York   Ninth Edition, 1982
   A concise but comprehensive summary of the principles of effective writing. Covers grammar, punctuation, spelling, usage, sentences, paragraphs,
   and complete compositions.

BOOKS ON USAGE

5. **A Dictionary of Contemporary**
   American Usage   Bergen and Cornelia Evans
   Random House, New York   First Edition, 1957
   Authoritative guide to preferred usage in American English. This book,
   now in its 14th printing, is witty, entertaining, and extremely useful. We
   consider it a basic reference.

_____

*Softcover edition available (an asterisk next to the publisher's name indicates that this is
the publisher of the softcover edition).

6. ***A Dictionary of Modern English Usage**   H.W. Fowler
   Oxford University Press, New York/Oxford   Second Edition, 1965
   A noted reference work in the conservative manner. If you write for British or Commonwealth (except Canadian) readers, refer to this guide. It also has a section on "Americanisms."

7. **A Dictionary of American-English Usage**
   Based on Fowler's Modern English Usage   Margaret Nicholson
   Oxford University Press, New York   First Edition, 1957
   A middle-of-the-road authority, between Fowler and Evans. (Since it is based on Fowler, this dictionary is rather on the conservative side.)

8. ***The Elements of Style**   William Strunk Jr. and E.B. White
   Macmillan Publishing Co., New York   Third Edition, 1979
   Classic condensation of the principles of good writing in English. Belongs on every writer's desk.

9. ***The Careful Writer:**
   A Modern Guide to English Usage   Theodore M. Bernstein
   Atheneum, New York   First Paperback Edition, 1965
   An entertaining guide to American usage—plenty of humor enlivens its pages. A useful feature is the large number of entries specifying which prepositions can or should follow many troublesome verbs, nouns, and adjectives.

10. ***Modern American Usage:**
    A Guide   Wilson Follet (ed.: Jacques Barzun)
    *Hill and Wang, New York   First Edition, 1966
    A lexicon of problem words in English, with helpful notes on pronunciation and punctuation. In-depth treatment of usage problems does not include elementary ones like *affect/effect* and *lie/lay*.

11. **Dictionary of Problem Words and Expressions**   Harry Shaw
    McGraw-Hill, New York   First Edition, 1975
    Basic coverage of the most common problems, from *adverse/averse* to *would of*. Includes a list of 300 trite expressions to avoid, principal parts of the most troublesome verbs, and shorter lists of euphemisms and wordy expressions that can be replaced.

12. ***Geowriting: A Guide to Writing,**   Edited by Wendell Cochran,
    Editing and Printing in Earth Science   Peter Fenner, and Mary Hill
    American Geological Institute, Falls Church, Va. Third Edition, 1979
    A compact guide on preparing geoscience articles for publication. The sections on Organization and Personal Style are by Melba Jerry Murray.

STYLE MANUALS

These books focus on typographic style and format rather than on literary style. Use them for help with punctuation, capitalization, abbreviations, acronyms, personal titles, use of numbers, and so forth. Some of the newer editions (like the "Stylebook" of *U.S. News and World Report*) have entries on sexist terms and recent slang.

13. **New York Times Manual of Style and Usage** — Edited by Lewis Jordan
Times Books, New York — Revised Edition, 1976

14. **University of Chicago Press Manual of Style** — Twelfth Edition
University of Chicago Press, Chicago — Revised 1969

15. ***U.S. News and World Report Stylebook For Writers and Editors** — Edited by Turner Rose, et al.
Revised by D.E. Pollard & R.G. Smith
*U.S. News & World Report, Washington, DC — Fifth Edition, 1984

16. ***U.S. Government Printing Office Style Manual**
*Government Printing Office, Washington, D.C. — Second Edition, 1984

17. **SPE Publications Style Guide**
Society of Petroleum Engineers of AIME, Richardson, TX — First Edition, 1984

READABILITY FORMULAS

While there are many books on readability (and formulas for calculating it), these two are the best-known. They can help you not only to detect sentences that are difficult to read, but also to improve them.

18. **The Art of Readable Writing** — Rudolf Flesch
Harper and Brothers, New York — First Edition, 1949

19. **The Technique of Clear Writing** — Robert Gunning
McGraw-Hill, New York — Revised Edition, 1968

GOOD READING ABOUT ENGLISH

Some of these books (marked [A]) have alphabetized entries for ready reference, but by and large these are not reference books in the usual sense. All of them, nevertheless, have important things to say about the use and misuse of English, and they all make entertaining reading or browsing.

20. ***Watch Your Language** — Theodore M. Bernstein
*Atheneum, New York — Hardcover edition, 1958; Softcover, 1965

21. **Dos, Don'ts & Maybes of English Usage [A]**   Theodore M. Bernstein
    Times Books, New York   First Edition, 1977

22. **On Language [A]**   William Safire
    Times Books, New York   First Edition, 1980

23. **What's the Good Word? [A]**   William Safire
    Times Books, New York   First Edition, 1982

24. **Less Than Words Can Say**   Richard Mitchell
    Little, Brown and Company, Boston   First Edition, 1979

25. **The Graves of Academe**   Richard Mitchell
    Little, Brown and Company, Boston   First Edition, 1981

26. **The Leaning Tower of Babel**   Richard Mitchell
    Little, Brown and Company, Boston   First Edition, 1984

27. ***Strictly Speaking**   Edwin Newman
    *Warner Books, Inc., New York   First Edition, 1975

28. ***A Civil Tongue**   Edwin Newman
    *Warner Books Inc., New York   First Edition, 1977

Appendix A
Answers to Tests

Test 1

If you are really intuitive, you will have marked 11 sentences with an **F**. Only one sentence does not contain an outright error.

(1) This hospital expects an emergency patient to realize that they don't have all the answers.

Faulty pronoun reference. The <u>hospital</u> (singular) should not be referred to as <u>they</u> (plural).

(2) The printer was connected to the computer, but it was found to be defective.

Uncertain pronoun reference. Does "it" refer to the printer or the computer? Try rewriting this sentence to show clearly (a) that the printer was defective and (b) that the computer was defective.

(3) Heating costs were reduced by thicker insulation and lowering the thermostat.

Nonparallel construction. Coordinating conjunctions (*and, or, nor, for, but,* and sometimes *yet*) join elements of equal rank and of similar construction. *Insulation* in this sentence is a noun; *lowering* is a verb form (a gerund). The sentence could be recast as follows: (a)". . . by thicker <u>insulation</u> and lower thermostat <u>settings</u> . . ." or (b) by using ". . . <u>installing</u> . . . and <u>lowering</u>. . . ." It's still not a great sentence. Try using active verbs.

(4) The two solutions were injected into the model and the liquid withdrawn.

In the second clause, the subject and verb do not agree in number. In the first clause the subject, *solutions,* is plural; its verb, *were injected,* is also plural (by virtue of the auxiliary *were*). The writer

259

assumed that this auxiliary would also serve the verb, *withdrawn*, in the second clause, but it won't. The singular subject of this clause, *liquid*, requires a singular auxiliary: *was* withdrawn.

(5) The process has the capacity of atmospheric pollution reduction.

This sentence has no outright grammatical errors, but it is needlessly wordy because of a distorted verb, *reduction*. What it means is, "The process reduces atmospheric pollution."

(6) By enlarging this annular passage, the controlling valve can rotate relative to its housing.

The prepositional phrase, "By enlarging this annular passage," dangles. It has nothing to modify. The controlling valve did not enlarge the annular passage. The housing certainly didn't do it. And ". . . can rotate by enlarging this . . . passage" would not make sense. How about "This enlarged annular passage allows the controlling valve to rotate."? Or "When this annular passage was enlarged, the controlling valve could rotate." Or "We enlarged this annular passage so the controlling valve could rotate."

(7) A spokesman said Deep South Airlines is as safe, if not safer, than it has ever been.

Incorrect comparison: *as safe* must be completed by *as*, not *than*: ". . . as safe as, if not safer than, . . ."

(8) Reduction in death and serious injury rates occur as a result of this change.

Subject and verb do not agree in number. The subject is *reduction,* not *rates*. Therefore the verb should be *occurs,* not *occur*. We can go further and point out that the sentence is weakened by a distorted verb. Instead of "Reduction in death and serious injury rates occurs . . ." why not "Death and serious injury rates are reduced . . ."? For that matter, why not make the passive verb active: "This change reduces death and serious-injury rates."

(9) If a high energy had existed it would tend to have broken up the solids.

A problem with the perfect tenses: ". . . it would tend to have broken up . . ." should be ". . . it would have tended to break up . . ." (or ". . . it probably would have broken up . . .").

(10) This modification is important in processing, and we have designed a parts kit, also.

Unrelated ideas. The two independent clauses joined by *and* are unrelated and do not belong in the same sentence. Moreover, the use of *also* wrongly suggests that designing a kit is an addition to the importance of the modification.

(11) In fact, a person has hurt themself on this equipment in the last 24 hours.

Faulty pronoun: there is no such word as *themself. Themselves* (plural) would also be incorrect for referring to "a person" (singular). Use *himself* unless the person is known to be female, in which case *herself* would be used. Or recast the sentence: "In fact, a person has been hurt using this equipment . . ."

(12) This device is used to reduce the scale and in reading the values.

Nonparallel construction—a very common error. Here the coordinating conjunction *and* incorrectly joins an infinitive *(to reduce)* and a prepositional phrase *(in reading)*. Make both phrases either infinitive phrases or prepositional phrases.

Test 2

IDENTIFYING VERBS (p. 174)

(1) Improvement of the 25-foot-wide dock <u>included</u> the installation of a new fendering system.
(2) Now the largest ships anticipated <u>can</u> <u>be</u> <u>accomodated</u> without overstressing the dock.
(3) This system <u>extends</u> the full length of the 1,208-foot-long dock, long enough to berth two 745-foot vessels simultaneously.

(4) New pile clusters, one upstream and the other downstream, <u>are</u> <u>used</u> for mooring lines.

(5) Storage and subsequent delivery, however, <u>can</u> <u>present</u> problems, especially in very cold weather.

(6) To provide storage, six 72,000-bbl tanks <u>were</u> <u>converted</u> by adding steam coils and insulation.

(7) All <u>are</u> cone-bottom tanks so that they <u>can</u> <u>be</u> completely <u>emptied</u>.

(8) Coils in each tank <u>total</u> approximately 6,500 feet of 4-inch pipe.

(9) A Varek tank-gauging system <u>is</u> <u>used</u> to keep the terminal operator informed of the contents and temperature of each tank.

(10) The boilers <u>produce</u> a substantial volume of 15-lb maximum steam, <u>which</u> <u>is</u> <u>circulated</u> through the tanks, and which <u>maintains</u> a temperature of 115° to 155°F.

Test 3

PICKING THE SKELETON (p. 175)

(1) The <u>head</u> produced by a particular pump <u>is</u> the <u>height</u> reached by the pumped liquid.

(2) The <u>head</u> of a pump <u>is</u> normally <u>given</u> in feet.

(3) From the topic on pressure, <u>you</u> <u>will</u> <u>remember</u> [that there <u>is</u> a relationship between the height of a column of liquid, its specific gravity, and the pressure exerted at the bottom of the column].

(4) Therefore, the <u>head</u> produced by a particular pump under certain operating conditions <u>is</u> an <u>indication</u> of the pressure (psig) developed by the pump.

(5) The <u>location</u> of the pump and the <u>level</u> of the liquid to be pumped <u>determine</u> the maximum <u>height</u> to be reached by the liquid. (Note: This sentence has a compound subject joined by the conjunction <u>and</u>.)

(6) To fully understand this subject of head, (you) look at the following pumping systems. (look at is the equivalent of see or observe)

(7) (You) assume in all these systems [that the pumping is being done by the same pump], and [that the liquid being pumped is water].

(8) (You) note in Fig. 2 [that water rises to the same level in the tank and pump].

(9) In this case pumping is raising the liquid 100 feet.

(10) Therefore we say [that the discharge head developed by the pump at a capacity of 100 gpm is 100 feet].

Test 4

DISTINGUISHING ACTIVE FROM PASSIVE VERBS (p. 176)

(1) Business leaders could hardly believe (A) their good fortune.

(2) They were offered (P) a facility near the downtown hub.

(3) Headed by Julius Jamieson, later a Justice of the State Supreme Court, a committee studied (A) the advisability of accepting the gift.

(4) It was suggested (P) that the name of the facility be changed (P) to honor him.

(5) A century after the name was selected (P), the choice had (A) unexpected consequences.

(6) Halley's comet is sometimes called (P) the "once-in-a-lifetime" comet.

(7) The lifespans of relatively few people coincide (I) with two passes of the comet around the sun.

(8) Halley's comet was expected (P) to be visible through binoculars and amateur telescopes.

(9) The comet was (I) visible from Earth in 1910.

(10) It is said (P) that frightened people bought (A) comet insurance and comet pills in case something terrible happened (I).

Test 5

VERBS AND VERBALS (p. 177)

(1) Constant lobbying to prevent damaging legislation is credited with influencing dinsinterested legislators. [**Complete**]

(2) In evaluating a merger, the fact that an industry is ailing or a company is failing. [**Incomplete**—There is no verb for the subject, fact.]

(3) Discovery of the genetic code grew out of having a theory different from accepted wisdom. [**Complete**]

(4) More hiring in companies producing new products and expanding existing lines and thus boosting employment. [**Incomplete**—There is no verb for the subject, hiring.]

(5) The greatest problem will be developing a reasonable explanation for this phenomenon. [**Complete**]

(6) Not buying insurance to cover medical costs that cannot be predicted and that are unlikely to be covered adequately. [**Incomplete**—No verb for the subject, (not) buying.]

(7) Managing your money is the subject of the next lecture. [**Complete**]

(8) Before proceeding with the next step, to enclose the sample in a coating of plastic. [**Incomplete**—This is a pair of phrases. There is no main clause with a subject and verb.]

Appendix B _____
The Murray System Applied to a Complex Report:

A FORMAL PROPOSAL TO A GOVERNMENT

The Situation:

The author, reservoir engineer Sam Wilson, worked for an American oil company, Megabux (Petrolandia) Inc., which was preparing to start another project in the country of Petrolandia under a government contract to increase recovery from an old oilfield by injection of water. Sam's report had to be addressed to Petrolandia's Minister of Energy (who had minimal knowledge of the technology involved) and sent to the attention of the Commissioner of Petroleum Contracts (who had a knowledgeable technical staff).

Under the terms of the Megabux contract, an engineering feasibility study with estimates of reserves and recoveries had been completed, and the output had now to be submitted to the government in the form of a detailed technical report. Although Sam was listed as the author of his own report, the transmittal letter was to be signed by his division manager. Before the manager signed, the report had to be approved by Sam's section head, as well as four other department managers. Moreover, Sam was anticipating some special problems in composing the report.

Step 1 — Analysis of People Involved

(1) WHAT does author Sam Wilson want?

Approval by the Petrolandia Minister of Energy to conduct two pilot waterfloods in the Djinn Lake Field.

WHAT does author offer?

Recovery of an additional 550 million bbl of oil from the Djinn Lake Field, with the Government's effective share amounting to 320 million bbl, for total Government revenues (including taxes) of $10.25 billion over the 12-year life of the project.

265

(2) WHAT are the readers' interests?

Reader	Interests
TO: The Hon. Minchak Merno Minister of Energy 610 Bureaucracy Tower Capitalville, Petrolandia	What's in it for us? How much oil can be produced commercially? How much will our share be? Will project create new jobs for Petrolandians? Will it create environmental problems? When can you begin? When can we expect results?
THROUGH: Mr. Lendak Nosei Commissioner of Petroleum Contracts and	Same as Minister's, plus: How reliable are your reserves and economic estimates? Is the plan technically feasible? Why two phases in the pilot project? Does the plan conform to the terms of the contract?
Nosei's Technical Staff, headed by Dr. Alno Terpak	Same as Minister's and Commissioner's, plus: How did you estimate recoverable oil? How did you calculate Government's share? How will you carry out the pilot waterflood? How good are the studies on which the plan was based? How did you determine rock and fluid properties? Reservoir geometry?
FROM: Henry W. Johnstone Manager, Exploration and Production Division Megabux (Petrolandia), Inc.	Same as Minister's and Commissioner's, plus: Overview of planned project; assurance of technical accuracy; correct handling of possible problems with Government (two-phase pilot project). Also: What's the expected cash impairment, year

by year? What's the pay-out period? What are the estimated recoverable reserves, risk-weighted and "unrisked"? What are the principal technical or geologic risks? How much will the pilot project reduce the risks?

THROUGH (internal):
  Goode Fillup
  Sam's Section Head

What's the latest reserve estimate for the Holi dolomite? What method did you use to calculate it? Does the Fatt sandstone reserve estimate reflect the new data from Geologic Research? Have you marshalled convincing arguments on technical feasibility of the main project and the two pilots? Does the report reflect well on our section? Will it satisfy our Department Manager?

  G.O. Ratio
  Sam's Department Manager

Are reserve calculations as reliable as we can manage? Is the DCF analysis correct and fully documented? Will the Government technocrats be satisfied with the accuracy of the engineering data? Will they be satisfied with the report itself? Will our Division Manager be satisfied with it?

  Manager,
  Production Department

What do we have to do? When? How many workovers? What plans for treatment of injection water?

  Manager,
  Economics & Planning Dept.

Have you accurately stated the economics we supplied?

| Manager,
Legal Department | Does the proposal conform to our contract with Petrolandia? Is there any wording in the report that could cause problems later? |

COPIES TO:

Manager, Drilling Department	How many injection wells are to be drilled? How deep? When?
Manager, Exploration Department	Is the geologic information we furnished treated accurately in the report? What additional support do you need from us during both the pilot phase and the full-scale project?
Manager, Purchasing Department	When will we have to start buying special pumps and other equipment? Do we have to get three bids, as we did on the last Petrolandia project?

(3) Do you foresee any special problems in writing this report?

Yes—I have three problems:

(a) The Petrolandia Minister of Energy and the Commissioner both speak English as a second language and prefer simple, direct language. Further, because of a new Government emphasis on economy, they are suspicious of "gold-plating." Division Manager Johnstone, however, in dealing with government bureaucrats, favors thick, elegant, and highly technical documents, because he believes that bureaucrats are impressed by ornate reports with an excess of detail and lots of technical jargon.

Strategy: (i) Get Dept. Manager Ratio (who is close to the Petrolandia Commissioner) to convince Mr. Johnstone that we must keep this proposal as short and simple as possible or risk rejection. (ii) Write a cover letter to the Energy Minister as a summary, replying to his special interests in a simple fashion.

(b) We will actually need two pilot projects: one for the Fatt sandstone reservoir and one for the Holi dolomite. But the Petrolandians expect only one pilot flood and may object, suspecting delays and overengineering.

Strategy: (i) Call it a "dual-phase pilot project" rather than two separate projects; (ii) lay out convincing reasons for the two phases (differences in reservoir characteristics); (iii) assure the Government that the two phases will be concurrent.

(c) Many of Mr. Johnstone's interests are company confidential or are outside the scope of this report, but he will want answers before he releases the proposal. There is no time to write a special report for Mr. J if I'm to get this report out on its due date (we're already getting close). Mr. J will also want the project manual.

Strategy: (i) Write a cover memo to Mr. J, giving summary replies to his questions and promising details as soon as the report is out. (ii) Later, use the report to the Government as the basis for a company confidential report. (iii) Expand the Government report into the project manual.

Steps 2*, 3, and 4 (*Combined here*)

The Essential Message	The Topic Outline	Illustrations
WHAT is the NEWS?		
Megabux (Petrolandia) Inc. requests approval to carry out a dual-phase pilot waterflood project in the Djinn Lake Field, Coastal Province, as detailed in this report.	I. Proposal for Pilot Waterflood	Fig. 1: Map

---

*See p. 277 for the cover letter based on Step 2.

WHY?

(1) This project will confirm our estimates of reserves and revenues and will provide the engineering parameters to be employed in designing and controlling the waterflood.

(2) Our economic analysis shows that the Government's total revenues over the 12-year life of the project will be $10.25 billion, including taxes. This estimate is based on feasibility studies demonstrating that 550 million bbl of secondary oil can be recovered by waterflooding the two main reservoirs of the Djinn Lake Field; the Government's effective share will be 320 million bbl.

II. Pilot Objectives

III. Petrolandia Revenue: $10.25 billion

A. Expected Daily Producing Rate

B. Cost Forecast

C. Assumed Market Price

D. Income from Taxes

IV. Secondary Reserves— 550 million bbl

A. Fatt Sandstone

B. Holi Dolomite

Table I: Engineering Criteria

Table II: Summary of Economic Yardsticks

Table III: Economic Calculations

Fig. 2: Generalized Stratigraphy

Table IV: Reserve Calculations

Table V: Reserve Calculations

HOW (will you conduct the program)?

(1) Because the petrophysical and fluid characteristics of the Holi dolomite and Fatt sandstone reservoirs are entirely different, the pilot project will have to be carried out in simultaneous phases—one for each reservoir.

V. The Pilot Program

Appendix: Work Program

(2) The Fatt sandstone phase of the pilot will be a double 5-spot in the northwest part of the field, requiring the drilling of two injector wells to around 2,300 meters each.

A. The Fatt Sandstone Phase

Figs. 3–7: Geologic Maps and Sections for Fatt Sandstone

(3) The Holi dolomite phase of the pilot project will involve a modified 5-spot in the southwest part of the field, with the conversion of one existing well (now shut in) to an injector.

B. The Holi Dolomite Phase

Figs. 8–12: Geologic Maps and Sections for Holi Dolomite

(4) In addition to pressure and fluid-volume measurements, we plan to use radioactive tracers in both phases of the pilot project to monitor the directions and rates of flood response.

C. Flood Response Monitoring

Table VI: Radioactive Tracer Characteristics

Figs. 13a–13d: Views of New Tracer Eqpt.

Table VII: Pressure Measurement Requirements

HOW (do you know this project is feasible)?

The pilot flood is based on a year-long feasibility study, including detailed core analyses and petrographic studies of cores by our research lab in Dallas, Texas.	D. Reservoir Studies	Figs. 14–16: Fatt ss.
	1. Fatt Sandstone Characteristics	Figs. 17, 18: Porosity-Permeability Relationships
		Table VIII: Imbibition Values
		Figs. 19–25: Porosity Distribution in Thin Section
	2. Holi Dolomite Characteristics	Table IX: Effective Porosities
		Fig. 26: Permeability Variations
		Figs. 27–33: Micrographs of Peels

HOW (will this project affect the environment)?

Since subsurface waters (from two different aquifers) will provide floodwater for the project, there will no environmental problems of any kind.	E. Environmental Impact: Nil

HOW (about work for Petrolandians)?

According to the work program outlined in the appendix, the full-scale waterflood project is expected to employ around 15 Petrolandian professionals and 45 oilfield workers.

| | F. Petrolandian Manpower | Table X: Manpower Requirements |
| | | Appendix: Work Program |

NOW WHAT?

With your approval, drilling of the two NW injector wells and conversion of the SW injector can begin immediately. We expect definitive response in both phases before the end of the year.

| | VI. Timing of Pilot Project | Fig. 34: Work Schedule (PERT diagram or Gantt chart) |

Step 5—The Expanded Outline

I. PROPOSAL FOR PILOT WATERFLOOD (Fig.1)
   Megabux (Petrolandia) Inc. is requesting approval of the Minister of Energy to conduct a two-phase pilot waterflood in the Djinn Lake Field, in the Coastal Province.

II. PILOT OBJECTIVES (Table I)
   A pilot waterflood is required to (1) confirm the economic estimates, which are now based only on data from the laboratory research and analyses conducted by our Dallas Research Center; and (2) refine the engineering parameters that will be used in designing the full-scale waterflood (Table I).

III. PETROLANDIA REVENUES: $10.25 BILLION (Table II)
   Economic analyses based on a reserve estimate of 550 million bbl of recoverable oil (see Chapter IV) indicate that the Government's revenue over the 12-year life of the proposed project will be US$ 10.25 billion.
   A. Expected Daily Producing Rate
      The combined producing rate from the Fatt sandstone and Holi dolomite reservoirs is expected to peak at 270,000 bbl of oil per day.
   B. Cost Forecast
      Expected project costs are summarized in Table III.

C. Assumed Market Price
Project economics are based on an initial oil price of
US$XX per barrel, escalating at 2% per year for the first
five years, and 5% annually thereafter.

D. Income from Taxes
The Government's estimated income from taxes paid by
Megabux on profits from the project is US$XXX (Table
III).

IV. SECONDARY RESERVES: 550 MILLION BARRELS (Fig.2)
The feasibility study carried out by our Dallas Research Center shows that 550 million bbl of crude oil can be recovered
from the old Djinn Lake Field by waterflooding the two main
reservoirs—the Fatt sandstone and the Holi dolomite.

A. Fatt Sandstone Reserves: 360 Million Barrels (Table IV)
The calculations summarized in Table IV give estimated
waterflood reserves of 360 million bbl for the Fatt sandstone.

B. Holi Dolomite Reserves: 190 Million Barrels (Table V)
Using the assumptions given in Table V, calculations for
the Holi dolomite show that a successful waterflood will
recover 190 million bbl of secondary oil from this reservoir.

V. THE PILOT WATERFLOOD PROGRAM (Appendix)
Because the petrophysical and fluid characteristics of the two
Djinn Lake reservoirs are entirely different, the pilot project
will be carried out in two phases, with a waterflood for each
reservoir. To save time, these will be conducted simultaneously in different sectors of the field. If results are not conclusive, we may want to expand one or the other of the pilot
floods.

A. Fatt Sandstone Pilot Phase (Figs. 3–7)
The Fatt sandstone phase of the pilot flood will involve a
double 5-spot in the northwest part of the field. Two injector wells will be drilled to around 2,300 meters each.

B. Holi Dolomite Pilot Phase (Figs. 8–12)
For the Holi dolomite phase of the project we have selected a 5-spot in the southwest part of the field. It will
require converting shut-in Well #118 into a water injector
and drilling five injection wells.

C. Flood Response Monitoring (Tables VI, VII; Figs. 13a–
13d)
In addition to up-to-date techniques for measuring pressure changes and fluid volumes during the pilot flood, we

plan to try the new radioactive tracer system illustrated in Fig. 13.

D. Reservoir Studies on Waterflood Feasibility

The one-year study that confirmed the feasibility of flooding the Djinn Lake reservoirs was based largely on the laboratory research and analyses carried out by our Dallas Research Center.

1. *Fatt sandstone characteristics*

Special core analyses and thin-section petrography of the Fatt sandstone confirmed that this prolific oil reservoir has several unusual characteristics.

(a) Unusual permeability distribution (Figs. 14–16)

The inferred permeability distribution in the Fatt formation, which is most unusual for this type of massive quartzose sandstone, is shown in Figs. 14–16.

(b) Porosity-permeability relationships (Figs. 17–18)

Permeability of the Fatt sandstone does not increase in the expected manner as porosity increases.

(c) Imbibition values (Table VIII)

Imbibition response as measured in 38 Fatt sandstone samples is summarized in Table VIII.

(d) Uniform porosity distribution in thin-section (Figs. 19–22)

Petrographic study of more than 400 thin-sections confirms that porosity distribution in the Fatt formation is highly uniform throughout the field.

(e) Effect of silica cement variations (Figs. 23–25)

Variations in silica cementation have three different effects in the Fatt reservoir.

2. *Holi dolomite characteristics*

Although the Holi dolomite has been cored far less than the Fatt sandstone in this field, enough cores were available to make all the analyses required for the feasibility study.

(a) Effective porosity studies (Table IX)

Effective porosity is consistently less than total porosity in the Holi reservoir.

(b) Permeability variations in the dolomite (Fig. 26)

Permeability variations in the Holi dolomite appear to be controlled by the distribution of vugs.

(c) Imbibition tests prove difficult (Table VIII)

The reliability of results shown in Table VIII ranges from fair to poor, owing to the difficulty of running imbibition tests on the available samples.

(d) Peel studies confirm highly variable porosity distribution (Figs. 27–31)

The highly variable distribution of porosity inferred from core analysis was confirmed by the petrographic study of some 250 acetate peels.

(e) Distribution of vugs controlled by matrix lithology (Figs. 32–33)

Vug porosity in the Holi formation is now shown to be a function of the lithology.

(f) New peel technique for dolomites

Petrographic studies of the Holi formation were greatly aided by the use of a new technique for making acetate peels, developed at the Petrolandia Institute of Technology under a grant from Megabux (USA) Inc.

E. Environmental Impact: Nil

We plan to use subsurface brines from the Mordak and Nimchak aquifers as flood waters and to recycle the recovered water after separation from the oil. The radioactive tracer to be used during the pilot test has a half-life of only four months. Therefore, there will be no environmental risks at all.

F. Petrolandian Manpower (Table X)

During its 12-year life, the Djinn Lake waterflood project is expected to provide employment for some 15 Petrolandian professionals and 45 oilfield workers, over and above the national personnel now working with Megabux (Petrolandia) Inc.

VI. TIMING OF PILOT PROJECT (Fig. 34)

We propose to begin both phases of the pilot waterflood as soon as Ministry approval is received. By beginning the drilling and conversion of injector wells within the next 30 days, we should have definitive results on both reservoirs by year-end.

APPENDIX: WORK PROGRAM

Letter of Transmittal

The following cover letter is derived from the Essential Message by selecting the relevant paragraphs that answer questions from the person to whom the cover letter is addressed—the Minister of Energy (see Analysis of People, p. 266)—and then dictating the text (Step 7) from those paragraphs.

```
The Hon. Minchak Merno
Minister of Petroleum
610 Bureaucracy Tower
Capitalville, Petrolandia

          Ref.: Proposed Pilot Waterflood
                Djinn Lake Field, Coastal Province

Dear Mr. Minister:

Under the terms of our 1988 contract with the Government
of Petrolandia, Megabux (Petrolandia) Inc. hereby requests
Ministry approval to carry out a two-phase experimental
project for secondary recovery of oil in the Djinn Lake
Field. A complete work program for the project is detailed
in the attached technical report.

Our economic analysis shows that the Government's total
revenue over the 12-year life of the project will be
$10.25 billion, including taxes. This figure is based on
the results of our feasibility study, which confirmed that
approximately 550 million barrels of secondary oil can be
recovered economically by waterflooding the two main
reservoirs of the field. The Government's share of profit
oil is estimated at 320 million barrels.

The proposed pilot project will confirm these estimates
and provide the engineering data that are essential for
efficient design of the project. Altogether some 15
Petrolandian professionals and 45 oilfield workers would
be employed during different phases over the life of the
project. Because we plan to use subsurface water sources
for injection and to reinject produced waters, the project
is not expected to affect the environment.

With your approval the pilot waterflood can begin
immediately. We would expect definitive results by the end
of this year, allowing the full-scale project to get under
way not later than the middle of next year.
```

Our technical personnel are ready to meet with members of
Commissioner Nosei's staff at your earliest convenience to
discuss any aspects of the attached study.

Respectfully submitted,

Henry W. Johnstone
General Manager

The Final Version of the Report

What follows is greatly condensed from the final version of the
Megabux technical report to the Petrolandian Minister of Energy.

Title Page

```
                Technical Report 90–11
          PROPOSED PILOT WATERFLOOD PROJECT
          DJINN LAKE FIELD, COASTAL PROVINCE
               REPUBLIC OF PETROLANDIA




                     Presented to
                 The Ministry of Energy
                 Republic of Petrolandia




                         By

             Megabux (Petrolandia) Inc.
                  305 Liberty Plaza
              Capitalville, Petrolandia

                   January, 1990
```

I. PROPOSAL FOR PILOT WATERFLOOD

Megabux (Petrolandia) Inc. is requesting the approval of
the Minister of Energy to conduct a two-phase pilot
waterflood in the Djinn Lake Field, in Coastal Province
(Fig. 1). This pilot project . . .

II. PILOT OBJECTIVES

A pilot waterflood is required to (1) confirm the economic
estimates, which are now based only on data from the
laboratory research and analyses conducted by our Dallas
Research Center; and (2) refine the engineering parameters
that will be used in designing the full-scale waterflood.
As Table I shows, . . .

III. PETROLANDIA REVENUES: $10.25 BILLION

Economic analyses based on a reserve estimate of 550
million bbl of recoverable oil (see Chapter IV) indicate
that the Government's revenue over the 12-year life of the
proposed project will be US$10.25 billion. In Table III we
have included . . .

A. Expected Daily Producing Rate
   The combined producing rate from the Fatt sandstone and
   Holi dolomite reservoirs is expected to peak at 270,000
   bbl of oil per day. This estimate is derived from . . .

B. Cost Forecast
   Expected project costs, summarized in Table III, have
   been . . .

C. Assumed Market Price
   Project economics are based on an initial oil price of
   US$XX per barrel, escalating at 2% per year for the
   first five years and 5% annually thereafter. We have
   not attempted to . . .

D. Income from Taxes
   The Government's estimated income from taxes paid by
   Megabux on profits from the project is US$ YY (Table
   III). This figure takes into account . . .

IV. SECONDARY RESERVES: 550 MILLION BARRELS

The feasibility study carried out by our Dallas Research
Center shows that 550 million bbl of crude oil can be
recovered from the old Djinn Lake Field by waterflooding

the two main reservoirs—the Fatt sandstone and the Holi
dolomite. As Fig. 2 shows, . . .

A. Fatt Sandstone Reserves: 360 Million Barrels
   The calculations summarized in Table IV give estimated
   waterflood reserves of 360 million bbl for the Fatt
   sandstone. It is important to note that . . .

B. Holi Dolomite Reserves: 190 Million Barrels
   Using the assumptions given in Table V, calculations
   for the Holi dolomite show that a successful waterflood
   will recover 190 million bbl of secondary oil from this
   reservoir. In contrast to the Fatt sandstone, the Holi
   dolomite . . .

V. THE PILOT WATERFLOOD PROGRAM

Because the petrophysical and fluid characteristics of the
two Djinn Lake reservoirs are entirely different, the
pilot project will be carried out in two phases, with a
waterflood for each reservoir. To save time, these will be
conducted simultaneously in different sectors of the
field. If results are not conclusive, we may want to
expand one or the other of the pilot floods. As shown in
the proposed work program (Appendix), . . .

A. Fatt Sandstone Pilot Phase
   The Fatt sandstone phase of the pilot flood will
   involve a double 5-spot in the northwest part of the
   field. Six injector wells will be drilled to around
   2,300 meters each. Fig. 3 gives . . .

B. Holi Dolomite Pilot Phase
   For the Holi dolomite phase of the project we have
   selected a double 5-spot in the southwest part of the
   field. It will require converting shut-in Well #118
   (Fig. 8) into a water injector and drilling five
   injection wells. As indicated in Fig. 9, . . .

C. Flood Response Monitoring
   In addition to up-to-date techniques for measuring
   pressure changes and fluid volumes during the pilot
   flood, we plan to try the new radioactive tracer system
   illustrated in Fig. 13. This system was selected
   especially because of the characteristics of the tracer
   (Table VI); it . . .

D. Reservoir Studies on Waterflood Feasibility
The one-year study that confirmed the feasibility of
flooding the Djinn Lake reservoirs was based largely on
the laboratory research and analyses carried out by our
Dallas Research Center. However, it also included . . .

　1. *Fatt sandstone characteristics*
　　Special core analyses and thin-section petrography
　　of the Fatt sandstone confirmed that this prolific
　　oil reservoir has several unusual characteristics.
　　The most important of these are discussed below.

　　(a) Unusual permeability distribution
　　　　The inferred permeability distribution in
　　　　the Fatt formation, which is most unusual
　　　　for this type of massive quartzose
　　　　sandstone, is shown in Figs. 14-16. It is
　　　　clear from Fig. 14 that . . .

　　(b) Porosity-permeability relationships
　　　　Figs. 17 and 18 show conclusively that
　　　　permeability of the Fatt sandstone does not
　　　　increase in the expected manner as porosity
　　　　increases. Rather, ...

　　(c) Imbibition values
　　　　Imbibition response as measured in 38 Fatt
　　　　sandstone samples is summarized in Table
　　　　VIII. Of the 38 samples suitable for
　　　　measurement, . . .

　　(d) Uniform porosity distribution in thin section
　　　　Petrographic study of more than 400 thin
　　　　sections confirms that porosity distribution
　　　　in the Fatt formation is highly uniform
　　　　throughout the field. Figs. 19-22 are
　　　　typical examples of . . .

　　(e) Effect of silica cement variations (Figs. 23-
　　　　24)
　　　　Variations in silica cementation have three
　　　　different effects in the Fatt reservoir.
　　　　Fig. 23 shows two of these effects: . . .

　2. *Holi dolomite characteristics*
　　Although the Holi dolomite has been cored far less
　　than the Fatt sandstone in this field, enough
　　cores were available to make all the analyses

required for the feasibility study. Six aspects of
these analyses . . .

(a) Effective porosity studies
    Effective porosity is consistently less than
    total porosity in the Holi reservoir. Data
    summarized in Table IX . . .

(b) Permeability variations in the dolomite
    Permeability variations in the Holi dolomite
    (Fig. 26) appear to be controlled by the
    distribution of vugs. Large vugs . . .

(c) Imbibition tests prove difficult
    The reliability of results shown in Table
    VIII ranges from fair to poor, owing to the
    difficulty of running imbibition tests on
    the available samples. The biggest problem
    . . .

(d) Peel studies confirm highly variable porosity
    distribution
    The highly variable distribution of porosity
    inferred from core analysis was confirmed by
    the petrographic study of some 250 acetate
    peels. Figs. 27 and 28 show . . .

(e) Distribution of vugs controlled by matrix
    lithology
    Vug porosity in the Holi formation is now
    shown to be a function of the lithology. The
    micrographs in Figs. 32–33 clearly
    demonstrate . . .

(f) New peel technique for dolomites
    Petrographic studies of the Holi formation
    were greatly aided by the use of a new
    technique for making acetate peels,
    developed at the Petrolandia Institute of
    Technology under a grant from Megabux (USA)
    Inc. This technique . . .

E. Environmental Impact: Nil
   We plan to use subsurface brines from the Mordak and
   Nimchak aquifers as flood waters and to recycle the
   recovered water after separation from the oil. The

radioactive tracer to be used during the pilot test has
a half-life of only four months. Therefore, we
anticipate no risk to the environment. Furthermore,
. . .

F. Petrolandian Manpower
During its 12-year life, the Djinn Lake waterflood
project is expected to provide employment for some 15
Petrolandian professionals and 45 oilfield workers,
over and above the national personnel now working with
Megabux (Petrolandia) Inc. Manpower requirements,
summarized in Table X, . . .

VI. TIMING OF PILOT PROJECT

We propose to begin both phases of the pilot waterflood as
soon as Ministry approval is received. By beginning the
drilling and conversion of injector wells within the next
30 days, we should have definitive results on both
reservoirs by year-end. The work schedule (Fig. 34) is a
modified Gantt chart that . . .

[APPENDIX: WORK PROGRAM]

Index _____